D1549592

THE ARCHAEOLOGY OF CANALS

N.B. Mention or illustration in this book
of any canal, canal structure or canal artefact
does not imply any right of access,
nor that it is safe to visit it.

THE ARCHAEOLOGY OF CANALS

P. J. G. Ransom

WORLD'S WORK LTD

18 01 82
0539M

I   7.00
*   7.00

By the same author:

Holiday Cruising in Ireland
A guide to Irish Inland Waterways David & Charles

Railways Revived
An Account of Preserved Steam Railways Faber &
Faber

Waterways Restored
An Account of Restored Rivers and Canals Faber &
Faber

Your Book of Canals
Canals for eleven- to fourteen-year-olds Faber &
Faber

Copyright © 1979 by P.J.G. Ransom

Designed by David Gibbons

Published by World's Work Ltd
The Windmill Press, Kingswood, Tadworth,
Surrey

Printed in Great Britain by BAS Printers Limited
Over Wallop, Stockbridge, Hampshire

SBN 437 14400 3

CONTENTS

LIST OF ILLUSTRATIONS

Illustrations are by the author, except where acknowledged

Colour photographs

Black and white illustrations

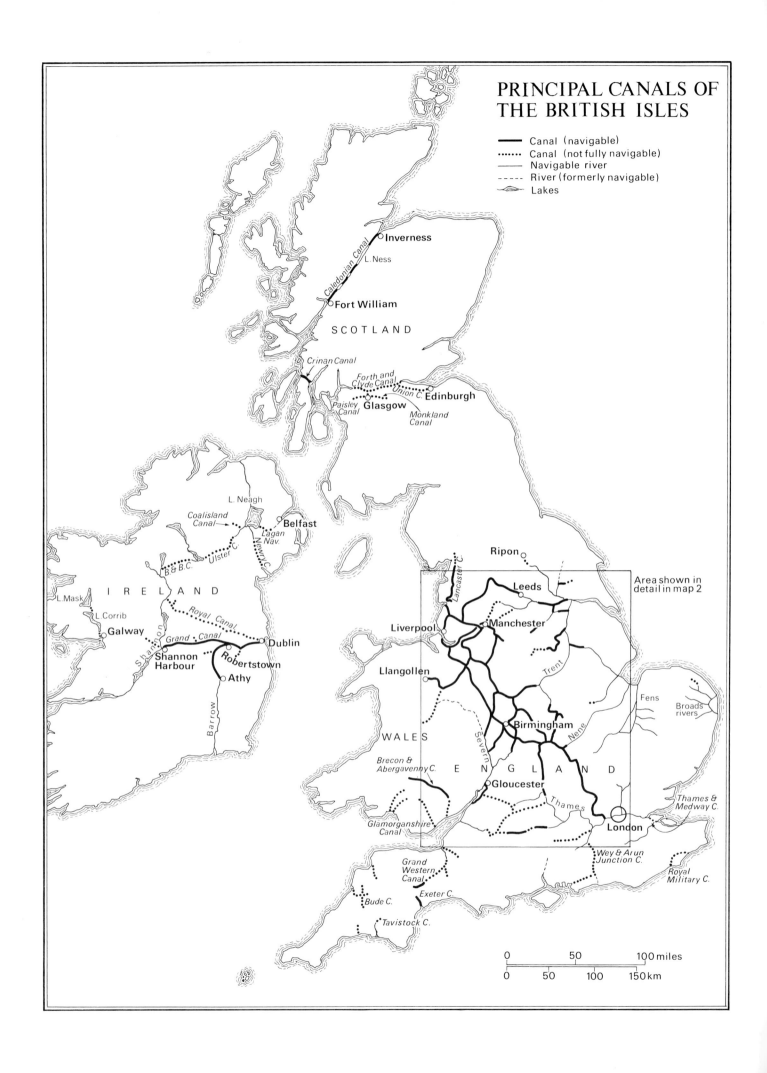

PRINCIPAL CANALS OF THE BRITISH ISLES

- ▬▬▬ Canal (navigable)
- ▪▪▪▪ Canal (not fully navigable)
- ───── Navigable river
- ----- River (formerly navigable)
- 〜〜 Lakes

Inverness
L. Ness
Caledonian Canal
Fort William

SCOTLAND

Crinan Canal
Forth and Clyde Canal
Union C. Edinburgh
Paisley Canal
Glasgow
Monkland Canal

L. Neagh
Coalisland Canal →
Belfast
Lagan Nav.
Ulster C.
Newry C.
B. & B.C.

IRELAND

L. Mask
L. Corrib
Galway
Royal Canal
Shannon
Grand Canal
Dublin
Shannon Harbour
Robertstown
Athy
Barrow

Ripon
Lancaster C.
Leeds
Liverpool
Manchester
Llangollen
Trent
Fens
Broads rivers
Birmingham
Nene

WALES

ENGLAND

Brecon & Abergavenny C.
Severn
Gloucester
Thames
Thames & Medway C.
Glamorganshire Canal
London
Wey & Arun Junction C.
Grand Western Canal
Royal Military C.
Bude C.
Exeter C.
Tavistock C.

Area shown in detail in map 2

| 0 | 50 | 100 miles |
| 0 | 50 100 | 150 km |

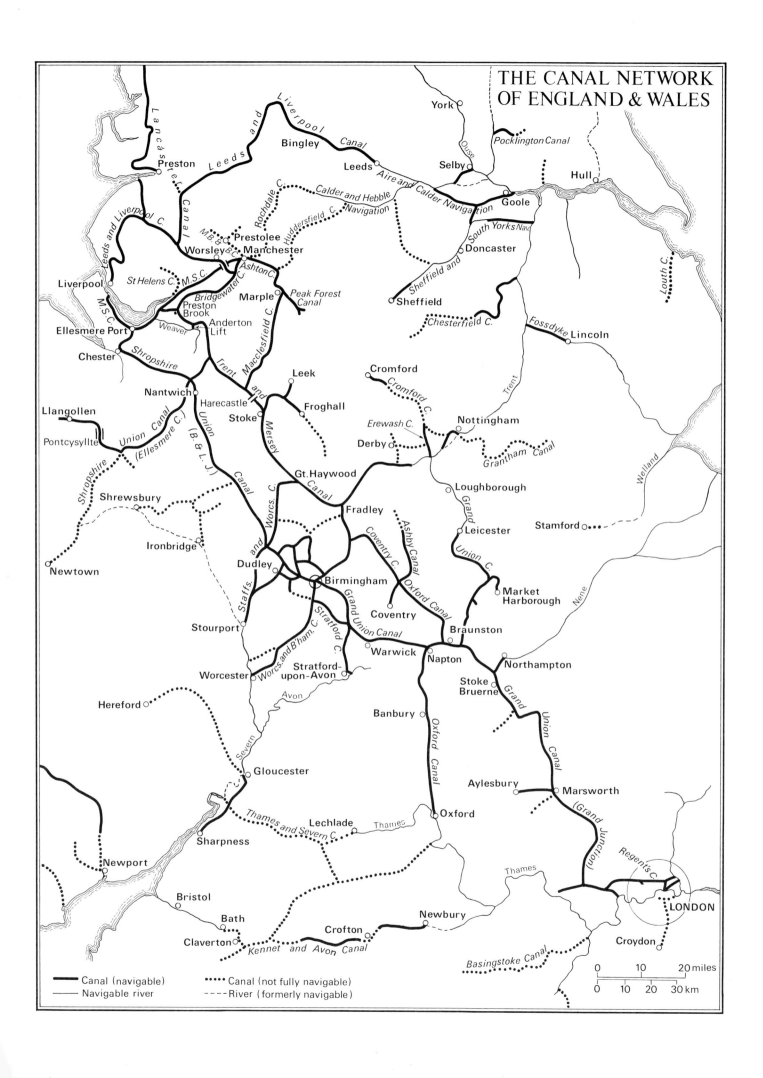

THE CANAL NETWORK
OF ENGLAND & WALES

York

Pocklington Canal

Selby

Hull

Preston

Leeds and Liverpool Canal

Bingley

Leeds

Aire and Calder Navigation

Goole

South Yorks Nav.

Lancaster Canal

Leeds and Liverpool C.

Rochdale C.

Calder and Hebble Navigation

Huddersfield C.

Prestolee

M.B.& B.C.

Worsley

Manchester

Ashton C.

Doncaster

Sheffield and

Louth C.

Liverpool

St Helens C.

M.S.C.

Bridgewater C.

Marple

Peak Forest Canal

Sheffield

Chesterfield C.

Fossdyke

Lincoln

Preston Brook

Anderton Lift

Macclesfield C.

Weaver

Ellesmere Port

Shropshire

Trent and Mersey

Chester

Leek

Cromford

Cromford C.

Trent

Nantwich

Harecastle

Froghall

Erewash C.

Nottingham

Llangollen

Union Canal

Stoke

Derby

Grantham Canal

Pontcysyllte

Shropshire Union Canal (Ellesmere C.)

Union Canal (B. & L. J.)

Gt. Haywood

Welland

Shrewsbury

Worcs. C.

Fradley

Loughborough

Grand Union C.

Stamford

Ironbridge

Dudley

Coventry C.

Ashby Canal

Leicester

Newtown

Staffs. and Worcs. C.

Birmingham

Oxford Canal

Market Harborough

Nene

Stourport

Stratford C.

Worcs. and B'ham. C.

Coventry

Grand Union Canal

Braunston

Northampton

Worcester

Stratford-upon-Avon

Warwick

Napton

Stoke Bruerne

Hereford

Avon

Banbury

Grand Union Canal

Severn

Aylesbury

Marsworth

Gloucester

Oxford Canal

(Grand Junction)

Thames and Severn C.

Lechlade

Thames

Oxford

Regent's C.

Sharpness

Thames

LONDON

Newport

Bristol

Croydon

Bath

Crofton

Newbury

Claverton

Kennet and Avon Canal

Basingstoke Canal

0   10   20 miles

0   10   20   30 km

——— Canal (navigable)   •••• Canal (not fully navigable)
——— Navigable river   - - - River (formerly navigable)

CHAPTER 1

SURVIVING TRACES OF THE PAST

Early and late canals

If I may commence by expressing a personal preference (in what is otherwise to be a strictly objective book), then it is for those canals built late in the canal era—during the 1820s and 1830s. These canals were promoted against a background of the gathering thunderclouds of competition by proposed railways, and when built they were superb engineering achievements. By that date the early canal builders' technique of meandering along the contours was old-fashioned: the late canals ran direct, by tall embankment and deep cutting, lengthy straight and sweeping curve.

Yet within a few years of their construction the storm broke, railways were all the rage and making these late canals was shown to have been in vain. Those who benefited from them most were the railway builders, who were able to inherit the administrative system of construction by contracting firms and the physical technique of cut-and-fill (making embankments with the spoil from cuttings) both of which canals had pioneered.

Soon, all canal transport seemed old-fashioned and canals in the British Isles sank into obscurity from which they have emerged only in the last thirty years. For, just as road transport was about to give canal transport its *coup de grâce*, there arose a public awareness both of the value of canals as an amenity and of their historical importance. During the period of a century or more, from the 1830s until the 1950s, canals had altered little, because on most canals the will to improve and the finance for improvements were equally lacking. There were, of course, a few shining exceptions—but the only big improvement to take place generally throughout the canals during that period was

the introduction of diesel power for boats to replace that of the horse. The canals remained, in the 1950s, to a large extent as they had been in the 1830s.

They had been built, most of them, during the eighty years or so previous to that, from 1759 onwards. Though this now seems a short period a long time ago, there had been great advances made in canal construction during the course of it: and the results of these, though they may not be obvious to the casual visitor, are clear to the experienced canaller. They may be seen, particularly, in the contrast when one passes from a canal built early in the canal era to one built late.

The Shropshire Union

Consider a cruise southwards along the main line of the Shropshire Union Canal, commencing on that section built as the Chester Canal during the years 1772–9. To be sure, there is a deep cutting through rock at Chester itself, a remarkable piece of engineering considering its early date. But little attempt seems to have been made to use the rock excavated from it: there is no corresponding embankment. Elsewhere the canal clings tenaciously to the surface of the ground, only in one place skipping nervously from one side of a shallow valley to the other by a low embankment, and crossing the little River Gowy by a small aqueduct. The method of following the contours and avoiding earthworks so far as possible was common among early engineers of British canals, such as James Brindley, and led in some places to exceptionally circuitous routes. Equally common was to consider each structure, each earthwork, in isolation from others. This is clear on the Chester Canal where the locks, for instance, are

placed at odd intervals according to variations in ground level.

How different is the continuation of the Shropshire Union south of Nantwich. This section was built in the late 1820s and 1830s as the Birmingham & Liverpool Junction Canal, filling a gap in the canal map to make a direct route between those places—and, in doing so, to thwart the plans of railway promoters. Shortly after leaving the Chester Canal, the B & LJ crosses over a main road by a handsome aqueduct built of cast iron (a material first used for canal structures only in the 1790s); then it leads purposefully onward, scorning the contours, crossing valleys by majestic embankments and slicing through high ground by impressive cuttings. For ease of maintenance and operation, locks are mostly grouped in flights, several locks over a short distance; and through flights of locks, the even gradient and flowing alignment are maintained by means of cutting, embankment and gentle curve. Throughout, the structures—bridges, locks, aqueducts—show a uniformity of design. Unlike early canals, this one was designed as a whole. Yet, though uniform, the structures are neither ugly nor boring, for in the 1830s the utilitarian was still made simply elegant.

In one respect, though, the B & LJ was retrograde. The Chester Canal is a wide canal, with locks able to take vessels 14 feet 3 inches wide and nearly 80 feet long, yet the B & LJ takes only narrow boats, 6 feet 11 inches wide and 72 feet long. Narrow boats and narrow canals had been designed early in the canal era to minimise construction costs of long-distance inland canals, but those early canals, such as the Chester Canal, which connected with estuaries, were often built large enough to take vessels already in use there. That the thirty-ton cargo of a narrow boat was inconveniently small soon became apparent and there was a time, during the 1790s, when it looked as though many canals

would be built as wide canals, and narrow canals widened. But they were not, and by the 1830s it was clear that however highly developed the engineering of the late canals might be in other respects, there was no point in building them wider than necessary for narrow boats. The Birmingham & Liverpool Junction, for instance, was dependent on traffic to and from the narrow canals of Birmingham.

This dependence on the narrow boat (and other craft not much larger) was in modern times to hamstring canals, so far as their chances of survival as a transport network were concerned, as effectively as the railway system would have been, had it been built to a track gauge of 2 feet 6 inches instead of 4 feet 8½ inches. Only where very much larger craft, carrying 500 tons or more, can operate does canal transport remain competitive, in Britain as on the Continent and elsewhere.

Some definitions

This seems an appropriate point to insert a couple of definitions. A canal may be taken as an artificial waterway, a highway of which the surface is water and the traffic boats, barges or ships. Archaeology is defined by the *Concise Oxford Dictionary* as the 'study of antiquities'—of ancient relics, that is. For the student of classical archaeology, or even of much industrial archaeology, that tends to mean digging up the past. The canal archaeologist is more fortunate. Canals were for so long disregarded, so little modernised, that, when one is on or beside a canal, relics of the past are all around. And a canal which remains in use is even better than one which is closed to navigation: equipment of

1/1 (above) Features of a late canal: fine masonry and a deep cutting on the Macclesfield Canal at bridge no. 43 (grid reference SJ 925720). The influence of the horse is present – the towpath, before crossing the canal, passes beneath the bridge and then starts to spiral back up on to it. Like this, there was no need to detach tow-ropes from boats. The top of the wall accompanying the spiral slopes up from ground level to give tow-ropes a snag-free run.

1/2 (next page) At Tyrley (grid reference SJ 689326) smooth alignment of Telford's Birmingham & Liverpool Junction Canal is continued down a flight of locks by means of cuttings and embankments. The canal was built c1828.

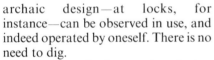

1/3 Surviving traces of the past. In 1978 a sign still indicates the Boat & Railway Hotel in Wharf Road, Stamford, Lincolnshire. Traffic on the Stamford Canal, which served the wharf, ceased in 1863.

1/4 (right) Narrow locks and a humped bridge typify the English canal scene. This is the Aylesbury branch of the Grand Junction Canal at Marsworth.

archaic design—at locks, for instance—can be observed in use, and indeed operated by oneself. There is no need to dig.

Not that digging archaeologists have not done good work on canals. They have cut through the cross-section of the Borrowstounness Canal, revealing the clay-puddled bed of that canal which was abandoned, incomplete, in 1796; they have uncovered parts of the private canal system which that remarkable eighteenth-century canal enthusiast Sir Roger Newdigate built to serve his estate by the Coventry Canal; and they have excavated a branch canal and its adjoining lime-kilns from beneath the debris of years to form a prominent component of the Black Country Museum.

But in general I have written this book in terms of surviving traces of the past. They are traces that may be seen by anyone cruising along a canal, or walking beside one, who knows what he or she is looking at—as on the Shropshire Union Canal just mentioned. So I have also written largely in terms of canals which remain open for

navigation; though not entirely for there are many important canal relics still to be seen along canals now closed.

From the very extensive network of canals which remains open, the most conspicuous absentee is the horse. Yet that cliché is not wholly appropriate, for we are today so unfamiliar with the needs of horses and horse-drawn boats that we no longer appreciate the extent to which the layout of canals and their immediate surroundings is still influenced by the former need to cater for horse-drawn traffic. Close to the typical canal junction, for instance, a bridge carries the towing path over one or other branch of the canal: to the steerer at the stern of a seventy-foot narrow boat, this bridge renders the junction totally and awkwardly blind. Yet in horse-drawn days it was no obstacle, for each boat was preceded on the bank by horse and driver. There are numerous other examples mentioned later—such as horse paths over tunnels, canal-side stables, and swing bridges, which swivel on the opposite side of the canal to the towpath, so as to offer no obstacle to tow-ropes.

Canals today

There are today in the British Isles about 1,630 miles of canals now navigable—that is about 1,400 in England, 30 in Wales, 90 in Scotland and 110 in Ireland. At their greatest extent, there were about 3,200 miles of canals in the British Isles.

The bulk of the British canal system, both navigable canals and those no longer so, is nationalised and belongs to the British Waterways Board. Some other canals still belong to canal companies established by Act of Parliament—notable among these is the Manchester Ship Canal Co., which inherited, through a complex process of canal history, the pioneer Bridgewater Canal. Other canals again belong to local authorities or the National Trust. In the Irish Republic most canals belong to Coras Iompair Eireann (CIE), the national transport company; their transfer to the Office of Public Works has been considered imminent for some years.

British Waterways Board's waterways (which include navigable rivers such as the Trent and the Severn) total about 2,000 miles and are divided by the Transport Act 1968 into: Commercial Waterways (about 300 miles), for commercial carriage of freight; Cruising Waterways (about 1,100 miles) for recreation; and 'the remainder' which the Board is obliged to treat in, with certain reservations, 'the most economical manner possible'. The Board has been under constant threat of reorganisation and/or dismemberment of its system since 1971, for its accounts show a chronic deficit. Since much of the canal system is still in use for pleasure traffic, it is worth briefly considering why.

The basic reason is this: that what the Board is maintaining is, except for the commercial waterways, the remains of an outmoded transport system. These remains, however, have no saleable value: the cost of eliminating them—draining canals, filling them in, making all their structures safe, etc—would be enormous. In 1965 BWB estimated that the cost to the Ex-

chequer of ruthlessly eliminating such canals or reducing them to minimal water channels would be equivalent to £600,000 a year. For another £300,000 to £350,000 a year they could be maintained in navigable order for pleasure cruising and amenity. The figures would now be much higher but there is no reason to suppose that the proportion has altered, and it does seem appropriate that a large part of the cost of maintaining cruising and remainder waterways should be met by the nation rather than the immediate users.

BWB has, however, been starved of finance by the Government—there is more about this at the end of chapter six—and the likely effect of this on the future maintenance of the Board's historic structures is alarming. There have already, in recent years, been many instances of sudden deterioration in the condition of tunnels and aqueducts, for instance, leading to long closures for repairs.

Nor does BWB have any general statutory duty towards maintenance of historic structures as such; although

water authorities do. Their duty, imposed by the Water Act 1973, is to have regard to the desirability of 'protecting buildings and other objects of architectural, archaeological or historic interest'. This is, perhaps, a consequence of proposals made at one stage to dismember the BWB system and distribute it among the water authorities. In practice, BWB on the whole treats its historic structures sympathetically.

Ancient monuments, listed buildings and conservation areas

What British Waterways Board does have are large numbers of structures which are scheduled as ancient monuments or listed as buildings of special architectural or historic interest. At the latest count, contained in the Board's annual report for 1977, it possessed eighty seven ancient monuments and 258 listed buildings.

Both 'monuments' and 'buildings' in this context include structures such as locks, bridges and aqueducts which are integral components of canals. Under various Acts of Parliament, ancient monuments are scheduled, and buildings of special architectural or historic interest are listed, by the Department of the Environment. The effect in both cases is similar, that they may not be demolished, nor repaired or altered in a way which would affect their character, without consent of the DOE or local authority.

In theory, this is excellent. In practice, applied to structures which form part of a working canal, it causes problems. There is no appeal against scheduling or listing, and these actions not only increase maintenance costs, because repairs have (quite rightly) to be in character, but also cause delay while notice is given that repairs are proposed—three months' notice in the case of ancient monuments. This requirement might well conflict with

BWB's obligation to keep canals open for traffic—if, for instance, an unsafe aqueduct, closed for urgent repairs, had to be kept closed for longer than would otherwise be necessary. So far as the extra costs are concerned, limited and discretionary Government and local authority grants are made, but not to BWB, at least on a regular basis.

Some canals pass through conservation areas, where the situation is even odder. In principle it seems excellent that the character and appearance of areas of special architectural and historic interest should be conserved, and that local authorities should designate such areas and treat the buildings within them much like listed buildings. However, owners need not be notified, and BWB seems not to know how much of its canal system is within conservation areas. It does, however, have some exemption from the resulting controls where operational structures are concerned.

There are to my knowledge at least six conservation areas based on lengths of canal: a fourteen-mile section of the Macclesfield Canal through Macclesfield, two long lengths of the Staffordshire & Worcestershire Canal, Wolverhampton locks on the Birmingham Canal Navigations, Shardlow canal village, and Fradley Junction where the Coventry Canal meets the Trent & Mersey. There are almost certainly others.

So while it is clearly desirable to ensure that the canals and their historic structures are fully protected and preserved, I am far from convinced that existing ancient monument, listed building and conservation area procedures are the best possible way to do so. Not only do they cause practical problems without offering solutions, but this piecemeal approach seems inappropriate for canal structures such as bridges and aqueducts which are in fact only part of a greater whole, the total canal. It is as though there were to be listed not a historic building but

only some of its features—a portico, a balustrade, say, but not the walls or roof.

All canal structures are interdependent, and what is needed is a total-canal approach. Further, since the British canals are unique, they justify unique treatment. A national conservation area covering the whole of the cruising waterways and most of the remainder is the sort of thing that is needed, with British Waterways Board not dismantled but reinforced. That means legislation to give it a duty similar to those of water authorities, to maintain its historic canals and their structures as such, and, where necessary, to provide funds for the purpose.

A note of explanation

Before concluding this introductory chapter I must mention that I take the word 'canals' in the title literally—this book is about canals rather than inland waterways and I include mention of navigable rivers and lakes only where they are relevant to canals. Similarly it is about the canals of the British Isles and I mention those of other countries only where they too have a definite relevance to those of these islands.

It is not possible, of course, to detail every canal relic. I have endeavoured to describe all the most important ones, and where those of less importance are mentioned, they may be taken as typical of others.

I have paid special attention to the Scottish and Irish canals, as they not only made some notable contributions to the history of canals in the British Isles, but they include remarkable relics not duplicated in England and Wales. Ireland in particular is a good place for the canal archaeologist, for the pace of life is slow and traces of the past survive longer than in Britain.

1/5 (right) Dundas Aqueduct on the Kennet & Avon Canal was the first canal structure to be scheduled as an ancient monument, as long ago as 1951. Dating from the middle of the canal era, the aqueduct itself is a fine piece of classical architecture, but the canal is still laid out to follow the contours and earthworks are kept to a minimum. To the left of the aqueduct was formerly the junction with the Somerset Coal Canal: traces of its course can be seen. Grid reference ST 784625.

CHAPTER 2

CANAL PRE-HISTORY

Roman canals

A field monument comparable in scale to Hadrian's Wall: that is how the Car Dyke was described in the report of a survey carried out under the auspices of the University of Nottingham in 1970. Or Offa's Dyke, or many Roman roads, it continued—yet the Car Dyke has received little attention from archaeologists.

There is no absolute proof that the Car Dyke, in Lincolnshire, and the Fossdyke are of Roman origin, but the Car Dyke in Cambridgeshire has been proved to be of Roman date, and mediaeval documents show that the two waterways in Lincolnshire were made before the Normans. The likelihood is that they formed a Roman waterway from the River Cam to the River Trent. The Car Dyke ran for 73 miles from the Cam, at Waterbeach north of Cambridge, to the Nene near Peterborough and onwards to the River Witham about 3 miles downstream from Lincoln; and the Fossdyke continued the line for 11 miles from the Witham at Lincoln to the Trent at Torksey. It is possible that they were built during the first century AD, more likely that they were built during the second. There were many Roman settlements along either side of the Lincolnshire Car Dyke; much Roman pottery has been found, and a bronze statuette of Mars was discovered in the bed of the Fossdyke at Torksey when it was cleaned out in the eighteenth century.

There is doubt, again, about the purpose of the Car Dyke: this may have been land drainage for it follows, in Lincolnshire, the 25 feet contour along the edge of the limestone escarpment and intercepts successive natural streams. But the scale of the works suggests that its primary purpose was navigation. During the survey already mentioned, made by Nottingham University's Department of Classical

and Archaeological Studies, the profile of the Car Dyke was examined at five locations in Lincolnshire by levelling across it. Conclusions were that in its original form the Car Dyke comprised three elements: a flat-bottomed trough with nearly vertical sides, a flat or gently sloping berm or ledge on one or both sides, and upcast banks on both sides. At one of the best preserved sites, grid reference TF 147392, the survey found upcast banks 4 feet 6 inches above the surrounding ground surface and 100 feet apart; the trough of the dyke was 33 feet wide at the top and 24 feet wide at the bottom. These dimensions are comparable with others made elsewhere on the dyke: they indicate a waterway large enough to be navigated by Roman river craft.

The remains of what may well have been a Roman riverside quay have been found at Lincoln, a nodal point in Roman communications. They comprise a 20 feet long stretch of large ashlar blocks, found at a depth of 16 feet and at a point about 100 yards north of the modern river channel. First century pottery has been found in the vicinity, and, south of the quay, a surface of clean sand and gravel which is thought to be the former river bed.

The Car Dyke went out of use after the Roman era, but its course is marked on 1:50,000 scale Ordnance Survey maps, and during a visit to south Lincolnshire in the late summer of 1978 I was able to examine it at several points. Near the village of Dyke, at grid reference TF 105224, I found it both impressive and remarkably well preserved, probably because the dyke here passes through grassland: no ploughs have eroded the upcast banks. This is one of the locations referred to in the Nottingham University survey, which reported that the dyke cut through an easterly projecting spur, the effect being to decrease the height of the western upcast bank which stands 1 foot 6 inches above the adjoining field

but 5 feet higher than land to the east. The eastern upcast bank has a height of no less than 6 feet, and the two banks are 105 feet apart. The projecting spur is scarcely visible to the naked eye, but the earthworks are indeed impressive, and suggest to the canaller a waterway of comparable scale to the Grand Junction Canal, if not larger. Deep down in roughly the centre of the dyke is a ditch, still in water, and looking very insignificant: the whole resembles the last vestiges of a closed canal, which of course is just what it is.

Such clear traces of the dyke appear to be unusual. As it approaches Dyke village, the western bank is first obscured by an orchard (of which the boundary hedge follows the ditch in the bottom of the dyke) and is then built over. The ditch, however, is more or less continuous along this part of the dyke—it has become incorporated into the land drainage system of the area. So it has been dredged, and the dredgings dumped to one side; and much of the adjoining land is plough. These two features have combined to obscure if not obliterate the original banks of the Roman dyke. When I visited its course at other locations in the vicinity—near Thurlby, Morton and Hacconby—I found the Car Dyke represented only by a small drainage channel which gave no indication of the scale or antiquity of the original.

The Middle Ages

The Fossdyke, like the Car Dyke, became neglected after the Roman era. In mediaeval times much use was made of rivers and the sea for transport, but while the River Witham and the East Coast roughly paralleled the Car Dyke, there was no alternative to the Fossdyke between Lincoln and the Trent. So in about the year 1121 King

2/1 Course of the Roman Car Dyke near the village of Dyke, Lincs (grid reference TF 105224). The upcast banks, here well preserved, are 105 feet apart, suggesting a canal intended for navigation.

Henry I authorised its re-opening: the Fossdyke was re-excavated, and by doing so a large area of land was drained, upon which new suburbs of Lincoln were built. Lincoln and Torksey became busy ports, but the dyke itself was ill-maintained and had to be reconstructed subsequently on several occasions. By 1365 for instance it was badly silted and the King by letters patent appointed commissioners to supervise its cleaning and repair.

During the twelfth century men elsewhere had begun to think of making artificial improvements to natural navigations. In Ireland, before 1178, a cut had been made across a low-lying island to provide a short route between the River Corrib near Galway and the lough of the same name. It was probably made by Franciscan friars: certainly it was, and is, called the Friars' Cut. It has been widened several times and still forms part of the Lough Corrib Navigation: I was able to pass along it while cruising there in 1969. Later in the twelfth century the Bishop of Winchester made the River Itchen navigable from Southampton.

The first canal built in Britain since Roman times was the Exeter Canal, opened in 1566 from the tidal Exe up to Exeter. The river itself had formerly been navigable up to that town, but had become obstructed by weirs and shallows. The canal was 1¾ miles long, had three locks and was built under an

Act of Parliament obtained by Exeter Corporation. With a depth of 3 feet, it carried lighters, into which ships lying in the estuary unloaded their cargoes.

The purpose of the Exeter Canal, then, was to bypass an unsatisfactory river: and it was with the improvement of rivers for navigation that the promoters of inland waterways in Britain occupied themselves for the next couple of centuries. During this period many side-cuts, or short artificial channels running roughly parallel to rivers and bypassing obstructions in them, were built, some of them several miles in length and justifying the description *lateral canal*. Improvement of navigable rivers in this way continued into the era of cross-country canals which followed. There was, however, a simpler way to improve natural rivers. The main drawbacks of large rivers for navigation were that they tended to flood after winter storms (about which little could be done) and to become too shallow during summer droughts. The latter condition could be ameliorated by dredging, by narrowing the channel so that it might scour itself out, and by building weirs to hold back the water. Smaller rivers, with comparatively

steep gradients, were too shallow and fast flowing to be navigable naturally, but they too could be deepened, and the current reduced, by construction of weirs at intervals.

In many places weirs already existed to provide a head of water for mills. What was needed was a means for barges to pass the weirs, and the simplest way to do so was by means of the structures known, in various parts of England, as flash locks, water gates, staunches or navigation weirs. The device was in effect a section of weir which could be opened, either by swinging it horizontally or raising it vertically, for barges to pass through. This was a simple but time-consuming operation, for before the flash lock could be opened, the level of the water in the reach above it had to be lowered until it was only a little higher than that in the reach below: and afterwards the upper reach had to be allowed to fill up again before navigation along it was possible. Therefore the pound lock, more expensive to build but quicker to use, gradually came into use. This is the type of lock which is familiar today. It can be visualised as having developed from construction of two flash locks very close together—no

2/2 The whole of the Exeter Canal is shown in this view, diverging from the River Exe and running roughly parallel as a lateral canal to enter the estuary in the distance. The nearest part of the canal is on the original 1566 course, except for the basin which, like the farthest parts of the canal, was a later addition.

more, in fact, than a boat's length apart—with sluices to raise or lower the level of water between them to equal that of the river above or below. Early pound locks on river navigations had sloping sides of turf: brick or masonry sides came later (they are essential on cross-country canals to avoid loss of water).

The Exeter Canal had pound locks, but for many years flash locks were the more common. The Earl of Arundel built them when making part of the River Arun navigable during the middle of the sixteenth century, and by 1600 there were about seventy of them on the Thames between London and Oxford. Improvements were needed between Abingdon and Oxford, however, and between 1624 and 1635 a commission appointed by Parliament built three pound locks on this section. These were some of the first pound locks to be built on an English river; all the same, navigation weirs did not disappear entirely from the Thames until the present century.

The middle years of the seventeenth century saw the draining of much of the Fens and, with this, the construction of many artificial watercourses. These were used for waterborne trade, but their chief purpose was land drainage and so they are outside the scope of this book. However, it is worth bearing in mind that so extensive a water-related project was being undertaken in the Fens when considering contemporary events elsewhere. Between 1636 and 1639 William Sandys made the Warwickshire Avon navigable from the Severn at Tewkesbury up to Stratford, having obtained letters patent from the King, and using a mixture of flash locks and pound locks, and between 1651 and 1653 the River Wey was made navigable from the Thames to Guildford under Act of Parliament.

In 1664 work started on the River Welland Navigation, for which an Act of Parliament had been passed as early as 1571. This resulted in construction of the longest canal so far built in Britain. As on the Exe, the problem was that the Welland, once navigable after a fashion from the sea to Stamford, Lincolnshire, had in its upper reaches become obstructed by watermills. The millers here were evidently not so accommodating as to allow their weirs to be adopted to hold up the water for navigation (as probably happened on the Warwickshire Avon, for instance), and an entirely new and artificial cut was made for six and three-quarter miles from Stamford down to Market Deeping. This lateral canal included eight locks: two more locks were built on the river downstream from Market Deeping. The whole was in use by the early 1670s.

The Canal du Midi

At this stage it is necessary to leave, temporarily, the history of British waterways to consider what was happening on the Continent: for while the Stamford Canal was the largest artificial waterway yet built in Britain, it pales into total insignificance when compared with the Canal du Midi then being built in France.

The French had borrowed the idea of building canals from the Italians, who had built them in North Italy during the fifteenth and sixteenth centuries—among the engineers was Leonardo da Vinci. In France, most of the Canal de Briare was built between 1604 and 1611: it was completed in 1642 after a hiatus which was political rather than technical in origin. This canal was over twenty-one miles long, had forty locks and provided a link between the River Seine and the River Loire east of Orléans. It was the first substantial cross-country, cross-watershed, canal.

To link the Mediterranean with the Atlantic had been a dream for many years: the success of the Canal de Briare prompted conversion of the dream into reality, by construction of the Canal du Midi. This was built large enough to take small seagoing vessels: even today its construction would be a big undertaking, in its own time it was the greatest civil engineering work in the world. The Canal du Midi was built between 1666 and 1681 by Pierre Paul Riquet and François Andreossy, under the patronage of Louis XIV, *le Roi Soleil*, and his finance minister Colbert. When complete it ran from the new port of Sète, on the Mediter-

2/3 *The very early Stamford Canal, built in the 1670s, was closed in 1863, but parts of it remain in water although, as here (grid reference TF 148096), overgrown with weed.*

ranean, by Béziers and Carcassonne to Toulouse, where it joined the River Garonne which was navigable to the Atlantic. It was no less than 149 miles long, between 56 feet and 90 feet wide at the surface and 6 feet deep, and had 101 locks, which carried it over a watershed 620 feet above sea level. Several of the locks were grouped in staircases, in which the top gates of one lock form the lower gates of the next. The lock chambers were of oval shape to withstand the pressure of the ground on either side; they are 19 feet 8 inches wide at the gates and 36 feet wide at the centre, and 100 feet long. An extremely extensive system of feeder channels was built to provide adequate water in an arid climate.

The Canal du Midi is still in use, commercial use, for freight-carrying barges; and large fleets of hire cruisers have been placed on it recently. Visiting the area in 1977 I was able to see two of its most famous features, the Malpas Tunnel which was the first canal tunnel in the world; and the eight-lock staircase at Fonséranes, the largest on the canal. Both of these are near Béziers and are marked on Michelin map sheet 83.

The west entrance of Malpas Tunnel was a remarkable sight. The tunnel was built on a grand scale: 24 feet wide and 19 feet from water level to the crown of the arch—dimensions which make the tunnels built much later by our own eighteenth-century canal engineers look like the holes of so many water rats. The tunnel is 180 yards long and has a towing path, though this is accessible at the western end only by steps, which suggests it was intended for vessels towed by men rather than horses. This western entrance has no portal but is formed in a wall of hugely pock-marked sandstone: were it not for the regular outline of the bore it would appear that the canal entered a vast natural cave.

Fonséranes locks I thought even more remarkable: a colossal staircase of eight locks which antedates the canal era in England by a century. Most of it was in use, though the lowest part of the staircase had been modified in a way which takes a moment to explain.

Some distance below the locks the canal originally entered the River Orb to cross it on the level, but in the 1850s a diversion was built to cross the river by an aqueduct. The water level of this diversion was that of the staircase's lowest-but-one lock chamber when full: therefore this chamber became the junction, rebuilt to become a triangular lock with an additional pair of gates leading to the new line of canal. The old line is disused, although the lowest part of the staircase leading to it appeared to be complete, and at the time of my visit the staircase had, in effect, six steps in full use. And as I walked up beside them to the top, a barge, spotless in green, red and shining black paint, started to work its way swiftly downwards—followed immediately by another equally smart.

Nothing to compare with the Fonséranes staircase was built in Britain until 130 years later, when the Caledonian Canal's staircase at Banavie was constructed. Now the Canal du Midi is being modernised and locks are being lengthened: it is intended to replace the Fonséranes staircase with a pair of very deep locks. I am glad to have seen the original historic structure in use.

Eventually, the Canal du Midi was to have a direct effect on canal construction in Britain. But not immediately—although it was famous and must have been well known. I am inclined to conjecture that King Charles II, a contemporary of its construction, may have had it in mind when he proposed a canal to enable warships to pass between the Firths of Forth and Clyde. Certainly when, a century later, firm proposals for the Forth & Clyde Canal were put forward, engineer John Smeaton likened it to the Canal du Midi.

It was of course usual that, long before individual rivers were made navigable and canals were built, proposals were being aired and ideas mooted, although it was seldom that the reigning monarch took part. Such proposals were far too numerous generally to mention here, although it is worth noting (since the rivers concerned have a prominent part to play later in the story) that in the 1660s a Bill to make the Rivers Mersey and Irwell navigable was turned down by Parliament, although later in the century the Mersey was made navigable up to Warrington.

By 1670 the Fossdyke and the River Witham were again 'silted, obstructed and in great decay' according to an Act of Parliament passed that year. The Act enabled the Mayor and Corporation of Lincoln to repair these waterways and appointed commissioners to oversee them. The Corporation does not seem to have been very successful and the Fossdyke remained barely navigable, until in 1740 it was leased to Richard Ellison who restored it and re-opened it. The Exeter Canal however was thoroughly dredged in the 1670s and extended for half a mile to join the river lower down; at the turn of the century it was enlarged again to take coasting and small seagoing craft.

Eighteenth century river navigations

The last years of the seventeenth century and the early part of the eighteenth were a time when there was great interest and activity in making rivers navigable. In 1699, after several earlier proposals, an Act of Parliament was passed for making the Rivers Aire and Calder navigable. The Aire & Calder Navigation was later to become the most important of British commercial waterways, but when first completed in 1704 it was on a scale which was modest compared with later developments. Its sixteen locks were probably about 60 feet long and 15 feet wide; the River Aire was made navigable from the tideway through Castleford up to Leeds and the River Calder from Castleford to Wakefield. The Act had appointed 'undertakers' from Leeds and Wakefield to carry out the work.

After that, waterways constructed or re-made came thick and fast. An Act passed in 1701 appointed undertakers to make the Yorkshire River Derwent navigable, and by about 1723 it was open from its confluence with the tidal Yorkshire Ouse up to Malton. The Itchen, which had been obstructed for a century, was re-opened about 1710. In 1712 Thomas Steers proposed to make both the River Douglas navigable from the Ribble estuary up to Wigan and the Rivers Mersey and Irwell navigable up to Manchester. It was some years before Acts were successfully obtained—in 1720 for the Douglas and 1721 for the Mersey & Irwell Navigation. The latter was completed first, being more or less usable by 1736; the Douglas was not completed until 1742, by which time Steers, as will shortly be recounted, was otherwise occupied in Ireland.

In 1712 and 1715 respectively came Acts for rivers which though at the time far removed from one another were later to be closely associated: the Bristol Avon and the Kennet. The Kennet was navigable from the Thames at Reading up to Newbury by 1723, and out of the 18½-mile length of the navigation, a total of 11½ miles was artificial cut. The Bristol Avon was opened up to Bath in 1727. In 1721, after many attempts, an Act was obtained to make the River Weaver navigable from the Mersey estuary up to Winsford in Cheshire; the work was done by 1732. It was also after many earlier attempts that an Act was obtained in 1726 to make the River Don navigable from Tinsley, near Sheffield, down to a point below Doncaster whence seventeenth-century drainage channels were navigable down to the Ouse estuary. The navigation was opened in 1752.

In the seventeenth and eighteenth centuries came development of waggonways, extremely primitive railways upon which wagons were pulled by horses over rails made of wood. In the long run they were to have serious adverse consequences for waterways, for they evolved into tramroads with iron rails and then into steam railways: but for the time being they and later tramroads helped by feeding waterways with traffic. They were used in north-east England to carry coal between pit and tidal Tyne; the first to serve a wholly inland waterway were probably those built between collieries and staithes, or quays, beside the Aire & Calder Navigation. One of them became the Middleton Railway, which has its own place in railway history (the first authorised by Act of Parliament, the first to use steam locomotives in commercial service . . . etc.).

Ireland

Here is it necessary to diverge once again from the mainstream of British waterway history, this time to consider what had been happening in Ireland. In mediaeval times, there as in England, natural waterways had been used as extensively as possible for navigation, mostly in small boats, although in Ireland natural waterways meant large lakes or loughs, as well as rivers, many of which flowed into or out of them. But it was not until the end of the seventeenth century that people began, to any great extent, to consider improving rivers to carry large craft; and then in 1715 came an Act of the Irish Parliament which envisaged not only the improvement of the principal rivers such as the Shannon, Boyne, Barrow and Erne, but also '. . . navigable and communicable passages for vessels of burthen to pass through . . . the . . . midland counties into the principal rivers . . .'. In other words, canals, such as were eventually built. The 1715 Act, however, had almost no practical result. It took two further Acts of Parliament, the second of which, in 1729, appointed Government commissioners responsible for inland navigation, before any substantial work was started, and then it took the form not of improvements to a river but of construction of a fully-fledged cross-watershed canal. This, the Newry Canal, was the first such canal in the British Isles.

Its main purpose was to carry coal: coal from mines which had been discovered in County Tyrone, to the west of Lough Neagh. It would be carried across the lough and up the Upper Bann River to Portadown, through the canal to Newry and thence by sea to Dublin. Work started on the canal in 1731; the first engineer was Richard Castle, a Huguenot refugee who had studied 'artificial navigation' on the Continent. At the end of 1736 however he was dismissed and replaced the following year by Thomas Steers from Lancashire. Steers completed the canal and it was opened in 1742: a notable achievement for its day, it was 18 miles long, about 45 feet wide and had fourteen locks.

Construction of the Coalisland Canal had been authorised in 1732. This canal was to lead from the River Blackwater (which flows into the south-west corner of Lough Neagh) to Coalisland, near the collieries. It proved very difficult to construct, across peat bog and quicksands, and though only 4½ miles long with seven locks (one of them a double lock* with two chambers) was not completed until 1787. In the meantime, an attempt to extend it by building a canal called Ducart's Canal, from Coalisland to one of the collieries, had proved a failure. The colliery lay on high ground: the canal was to rise not by locks but by inclined planes up which small boats would be drawn on cradles running on rails. These inclined planes, ahead of their time, did not work satisfactorily.

Ducart's Canal was built during the late 1760s and 1770s. Before this, much else had happened. A further Act of Parliament in 1751 had incorporated the commissioners into The Corporation for Promoting and Carrying on an Inland Navigation in Ireland, and then, in a political atmosphere which favoured Government expenditure on public works (and seems to have been accompanied by a certain amount of political and financial skulduggery!) work on several extensive inland waterways was started more or less simultaneously. These included: in 1755, the Shannon Navigation from Limerick to Carrick-on-Shannon (and later, beyond), a river-and-lough navigation with some extensive side canals, particularly between Limerick and Killaloe; in 1756 the Lagan Navigation from Belfast to Lough Neagh, partly river navigation and partly canal; in the same year the Grand Canal, wholly artificial, from Dublin to the Shannon; and in 1759 the Boyne and Barrow Navigations, both of them rivers eventually made navigable. The engineer for all these schemes was Thomas Omer. The emphasis is that work *started*: the only one of these projects to reach anything resembling completion during the present period was the Lagan Navigation, of which the first, navigable river, section was opened for about eleven miles from Belfast to Lisburn in 1763. The Grand Canal, however, was by far the most ambitious canal yet authorised in the British Isles; as planned it was to be a cross-country canal some seventy miles long to reach the Shannon near Banagher. Some of Omer's work on it was incorporated into the canal as

---

* Irish practice is to count a double lock, i.e., a staircase with two chambers, as one lock when totalling the number of locks on a waterway. No staircases with more than two chambers were built in Ireland. In Britain, where staircases with as many as eight chambers were built, the number of lock chambers is counted.

later built, and is described in chapter ten. A more northerly route to Lough Ree on the Shannon was also considered, proposals for which many years later evolved into the Royal Canal. At this period Omer was also engineer to the Newry Ship Canal, which opened in 1769 to enable larger vessels to reach Newry than had been able to do so by tidal river. It was later enlarged and extended.

The Sankey Brook

Thomas Steers, in addition to working on the Mersey & Irwell, the Douglas and the Newry Canal, was dock engineer at Liverpool. His pupil and successor there was Henry Berry, and when Liverpool merchants wished to make the Sankey Brook navigable, it was Berry who did the survey. The Sankey Brook led from the Mersey estuary northwards to St Helens, near which place there were coal-mines needing cheap transport to Liverpool. The lowest mile or so was tidal and

already navigable to Sankey Bridges near Warrington.

An Act of Parliament for making the 'River or Brook called the Sankey Brook' navigable was obtained in 1755 and included powers, as did similar Acts, to allow the undertakers to make new cuts and canals through land adjoining the river. In fact, the Sankey Brook was small, too small to be made navigable, and Berry used these powers to construct an entirely artificial lateral canal from Sankey Bridges to St Helens which was connected to the brook only in order to obtain a supply of water.

It is probable that Berry, who must

surely have learnt about the building of the Newry Canal first-hand from Steers, had intended a canal to St Helens from the start, but that since a Bill for a canal in the same district, from Salford to Wigan, had been thrown out by Parliament in 1754, familiar powers for a river navigation to St Helens were sought from Parliament the following year. At any rate, most of the St Helens Canal, as it came to be called, was open by 1757: 8 miles and ten locks including a staircase pair. Later, short branches were added, and successive south-westward extensions to easier entrances from the Mersey than the original.

2/4 (preceding page) The Canal du Midi, France. A barge descends the staircase at Fonséranes in 1977. This part of the Canal du Midi was built about 1679, eighty years before work started on the English canal system. The Duke of Bridgewater studied this canal during his education on the Continent, and works such as this must have inspired him.

2/5 (right) Locks on some navigations, where water was plentiful, had sloping sides of turf. This is Sheffield Lock on the Kennet Navigation (grid reference SU 649706). To prevent boats settling on the sloping sides while descending the lock, a framework has been built by the owning Great Western Railway from redundant bridge-section rails with which its railway was originally laid.

John Smeaton and the Calder & Hebble

In that same year of 1757 a survey was made of the River Calder. This was already navigable as far upstream as Wakefield, as part of the Aire & Calder Navigation; the intention now was to make it navigable further, together with a short length of the River Hebble, so as to reach Halifax. Similar schemes had been proposed before and about 1741 a Bill, with which Thomas Steers was associated, had been introduced to Parliament but was defeated. Halifax badly needed inward transport for raw wool, and in 1756 a committee was formed to seek another Act. It was at this committee's request that the new survey was made, by Yorkshireman John Smeaton.

John Smeaton is called the first British civil engineer. He was born, near Leeds, in 1724, and he was intended for his father's profession of attorney; but having great mechanical ability he abandoned his studies in order to make scientific instruments. In 1754 he travelled in the Low Countries studying canals, locks and harbours. It was not, however, through inland waterways that he made his name, although eventually he became associated with a great many, but by construction of the Eddystone Lighthouse off Plymouth. Two earlier lighthouses on the site had been destroyed, the first by a gale, the second by fire. Smeaton's was built of stone, unlike its predecessors; work started in 1756 and the lighthouse was completed in 1759. Since work on it could be done only in summer, Smeaton was able to survey the Calder in the autumn of 1757. It was his first waterway.

In 1758 an Act of Parliament was obtained; the western terminus had been moved further up the Calder to Sowerby Bridge, and the Act set up commissioners to run the navigation. (A later Act, in 1769, replaced them by the Company of Proprietors of the Calder & Hebble Navigation.) The work of construction started late in 1759, after the lighthouse had been completed. It took 5 years to finish the first 16 miles up to Brighouse.

A proposal—possibly a revival of an earlier one—was made in 1758 for a canal from the Trent to the Potteries. Its principal promoters were Earl Gower and Lord Anson and to make the survey they employed an engineer and millwright who was making a name for himself as a man of great ingenuity in his field: James Brindley. There was no immediate result, however, and Brindley and this canal belong properly to the next chapter.

A brief overall look at the state of waterways in England in 1759 is now worthwhile, for much was about to happen. Two great rivers, the Severn and the Trent, were naturally navigable, and much used for navigation. The Severn was totally unimproved, subject to interruptions from floods and drought, but vessels moved free of toll as far upstream as Welshpool. The Trent was in a similar state for much of its length, as far upstream as Wilden Ferry in Derbyshire; then for nineteen miles further upstream to Burton it had been made navigable at the end of the previous century. Elsewhere, river navigations had been made at a steadily increasing pace over the past 200 years: but they *were* all river navigations, or, in a few instances, lateral canals. Inland water transport was still based on rivers and river basins: nowhere in England did a canal yet cross a watershed.

CHAPTER 3

THE START OF THE CANAL ERA

The Duke of Bridgewater's Canal

In the late 1750s came the combination of men and circumstances which was to launch the canal era. Francis Egerton, third Duke of Bridgewater, a young man crossed in love, deserted the fashionable round in London to develop his estate at Worsley near Manchester. That was in 1759, but he had been working on plans for Worsley, one of many estates he owned, since coming into his inheritance in 1757 at the age of 21. He had found at Worsley an old-established coal-mine which could no longer be worked satisfactorily, for it was wanting in both drainage and transport.

The point was rubbed home by the opening in 1757 of the first part of the Sankey Brook Navigation, which placed the Worsley mine, dependent on inadequate road transport, at still greater disadvantage. Its traditional market was around Lymm in Cheshire, but this area was now open to competition from coal carried down the Sankey Brook.

The Duke was already well aware of the potential of canals. After an extremely unhappy childhood (which seems to have resulted in great determination of character) his education had culminated in the Grand Tour of Europe: during this he had made a point of visiting the Canal du Midi, and had examined it (travelling by road) from end to end. Subsequently he studied science and engineering at Lyons. Samuel Egerton, his favourite uncle, with whom he had spent much time at Tatton Hall, Cheshire, had made his fortune trading in Venice, no less, and his brother-in-law (and guardian until he attained his majority) Earl Gower, of Trentham near Stoke on Trent, was already

thinking in terms of the project which would eventually become the Trent & Mersey Canal.

There had already been two attempts to promote water transport to Worsley. As early as 1737, after a survey by Thomas Steers, the Mersey & Irwell Navigation had obtained an Act of Parliament enabling them to make the Worsley Brook navigable down to the Irwell. No action had resulted, probably because of the expense of making many locks. Then in 1753 and 1754 there had been the attempt to obtain an Act for a canal from Manchester to the Wigan area, which would have served Worsley. This failed because of opposition.

By 1758 the Duke himself was planning a canal from Worsley via Patricroft to a terminus at Salford, across the Irwell from Manchester. He was careful to meet the criticisms of the earlier Bill by, for instance, saying that water would be drawn from the Worsley Brook rather than from the Irwell above Manchester, which had been proposed previously and would have harmed that town.

In March 1759 he successfully defeated opposition to obtain an Act of Parliament enabling him to make in effect not one but two canals: from Worsley south-east to Salford, and from Worsley south-west to Hollins Ferry and a junction with the Mersey & Irwell Navigation. By these, coal could be carried both to Manchester and to the Lymm area.

Construction started the same year with John Gilbert, the Duke's agent for the Worsley estate, as resident engineer. Work commenced at Worsley on both the lines just mentioned, and also in a third, more remarkable direction: into the coal-mine itself. The usual way to drain a coal-mine located beneath high ground was to drive a sough or drainage tunnel to the lowest coal-seam from a point on low ground nearby. The sough at Worsley was inadequate and too high up and it was

John Gilbert's idea to drive a better and lower one which would have two important additional functions: by linking its outlet to the surface canal, water draining from the mine would be transformed from useless waste to an invaluable supply; and by making the sough of larger diameter than usual, boats from the canal could be taken directly inside the mine to be loaded. And this is what was done.

In fact two parallel tunnels were eventually dug, which joined together about 500 yards inside the hill; one tunnel was used for loaded boats leaving the mine and the other for empty boats returning to it.

John Gilbert's elder brother Thomas was general land agent and legal advisor both to the Duke and to Earl Gower. In 1758, as already mentioned, Gower and associates had been considering a canal from the Potteries to the Trent at Wilden Ferry and, to make the survey for this, John Gilbert had already introduced to the Earl a man called James Brindley.

Of all the names connected with canals in Britain, Brindley's is the best known. To him we owe the construction of a canal network. Yet in achieving this, he was to fuse together the ideas of many others. He was, as near as can be expressed in today's terms, a consulting engineer. His particular contributions lay in, firstly, great skill in laying out and designing watercourses and their artefacts, and, secondly, immense ability to promote convincingly the idea of canals before Parliamentary committees and indeed any other audience.

He was born in 1716 and in 1735 apprenticed to wheel- and millwright Abraham Bennett at Sutton, near Macclesfield. Bennett's workshop, though altered, still stands and bears a commemorative plaque. By happy coincidence, almost a century after Brindley's apprenticeship, the Macclesfield Canal, one of the last and finest to be built, was laid out to pass

3/1 The young Duke of Bridgewater demonstrates his and his engineers' achievements in carrying the Bridgewater Canal over a road and a navigable river at Barton. Gangs of men are bow-hauling 'flats' on the Mersey & Irwell Navigation, but horses are used for haulage on the new canal.

3/2 (overleaf) The original aqueduct across the river at Barton was demolished when the Manchester Ship Canal was built, and replaced by the Barton Swing Aqueduct, seen in the background of this picture. In the foreground are the remains of the stone-built embankment, with buttress, by which the original canal approached its aqueduct, and which can be seen in illustration 3/1 beyond the Duke's left coat pocket.

close by. Today's boaters, if they moor near Sutton Aqueduct, will find the workshop (attached to a private house) beside the road a hundred yards or so to the east (grid reference SJ 926717). From it, the aqueduct is prominently in view; it has separate arches at different levels for road and river and a masonry spillway, gracefully curving down the side of the embankment nearby, carries excess water down from the canal. Brindley, one feels, would have approved.

This is to jump ahead. Brindley was by modern standards semi-literate. Certainly his spelling was idiosyncratic: but spelling was less standardised then than now. This did not prevent his becoming an engineer, and a highly gifted one. In those days water-mills and windmills provided the power for every sort of industry; they were among the highest forms of technology known; and inherent in construction of water-mills was experience of watercourse engineering. From 1752 to 1756 Brindley was at work on a complex and successful scheme, which included a 600-yard tunnel, to use water power for mine drainage at Clifton, which is only about three miles from Worsley.

In 1759 Brindley was invited to Worsley itself and in July became 'consulting engineer' for the Duke's canal. The trio was complete: Bridgewater, Gilbert, Brindley.

One immediate effect was a change of plan. The canal was no longer to go to Salford: instead it was to cross over the Irwell at Barton and go, via Stretford, to Manchester itself. From Stretford a short branch was to run south west to Longford Bridge, whence coal would be supplied to places in Cheshire. Work on the Hollins Ferry line ceased. It seems likely that the Duke was already thinking in terms of making a canal in the direction of Liverpool by extending the Longford branch to the Mersey estuary, which would serve his Lymm markets en route and be wholly independent of the Mersey & Irwell Navigation.

To cross over the Mersey & Irwell Navigation at Barton, Brindley designed and built Barton Aqueduct. This was original work, for no precedent for such a structure existed in Britain and sceptics called it a castle in the air. The Duke, on the other hand, had no doubt seen aqueducts in use on

the Canal du Midi and would have been well aware that such things were entirely practicable. While the aqueduct was being built, in 1760, the canal was opened as far as Barton and coal transferred by crane to barges or 'flats' owned by the Duke on the river. In 1761 the aqueduct was completed and the canal opened across it as far as Longford, from which place coal for Manchester was carried by cart. The Barton Aqueduct, carrying canal boats over other vessels on the Mersey & Irwell Navigation, became popular with sightseers. It must have provided food for thought, as well: here for the first time in England was a canal totally independent of any river. The aqueduct emphasised the point.

The canal was eventually completed through to Manchester in 1765 and a link was made there with the Mersey & Irwell Navigation which enabled vessels to pass from one to the other. Before this, in 1762, the Duke had obtained an Act of Parliament to extend his canal from Longford Bridge to the Hempstones, about one and a half miles above Runcorn on the Mersey estuary, down which flats bound for Liverpool would continue. The first section of this line, as far as Altrincham, was opened in 1766.

The Trent & Mersey

In the meantime, developments had been taking place elsewhere which meant that the Bridgewater Canal was going to be the first component of a far greater network of canals. Brindley had made his survey for a canal from the Trent at Wilden Ferry to the Potteries in 1758, and John Smeaton in 1761 had suggested that this canal might be extended westwards beyond the Potteries to join a navigable river flowing into the 'western sea'—over the watershed, in other words, to provide a coast-to-coast water route across England. But the scheme lay fallow until taken up by Josiah Wedgwood, the noted potter, and various associates in 1765. The Potteries needed supplies of china clay, flints and salt (for glazing), and they also needed safe transport for their fragile finished wares. At that time they were dependent on pack horses and wagons

(running on very rough roads), as far as the nearest navigable rivers, the Trent at Burton, the Weaver and the Severn at Bewdley. The attractions of a waterway which would serve the Potteries themselves were clear, and construction of the Bridgewater Canal showed that it was practicable.

Wedgwood met Brindley and found him enthusiastic about reviving the scheme: before long an outline plan was produced for a canal seventy six miles long from Wilden Ferry to the Weaver at Frodsham, with a long branch to Lichfield and Birmingham. A canal to the Severn was suggested too.

In one respect the proposal was highly original: this was to be a narrow canal. Up till this time, waterways had

almost invariably been built to suit existing vessels on adjacent waterways. Such vessels were themselves often large enough to be suitable for estuary and coastal passages. This had resulted in locks generally between 50 and 90 feet long and between 12 and 17 feet wide. Channels, bridges and other structures had to be wide in proportion. The Duke of Bridgewater's canal was being built wide enough to take the sailing flats already in use on the Mersey & Irwell, the Mersey estuary and connecting waterways. For the Worsley mine, however, much smaller boats had inevitably to be used. They measured approximately 47 feet by $4\frac{1}{2}$ feet and it had been found convenient to use them not merely for transport underground, but also to

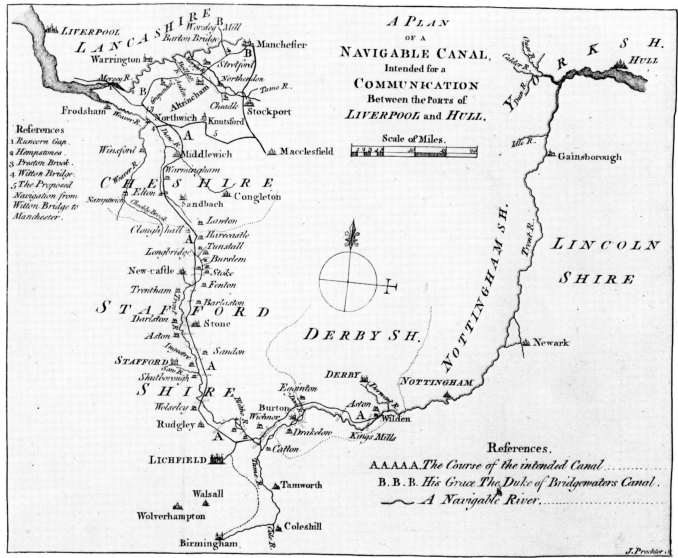

A PLAN
OF A
NAVIGABLE CANAL,
Intended for a
COMMUNICATION
Between the PORTS of
LIVERPOOL and HULL.

Scale of Miles.

References
1 Runcorn Gap.
2 Hempstones.
3 Preston Brook.
4 Witton Bridge.
5 The Proposed Navigation from Witton Bridge to Manchester.

References.
A.A.A.A.A. *The Course of the intended Canal*..........
B.B.B. *His Grace The Duke of Bridgewaters Canal*.
A Navigable River..........

J. Prockter.

3/3 (opposite) James Brindley. Barton Aqueduct appears in the background, seen from the east, the opposite side to the view in illustration 3/1.

3/4 (above) The Trent & Mersey Canal, and connections, are planned.

length) and thoroughly sound. Two hundred years later, in the 1960s, it was to prove crippling to canal trade throughout most of England, but one can hardly blame Brindley and his associates for failing to see so far ahead!

For a canal of its extent, the Duke of Bridgewater's Canal was unique in being the personal property of an individual, even though he financed it largely by borrowing (at its greatest, in 1787, the Duke's canal debt reached £346,371). There were thoughts of administering the Trent & Mersey through a trust in the public interest, comparable to a turnpike trust, but in due course the promoters decided to form a company, the *Company of Proprietors of the Navigation from the Trent to the Mersey*. *Trent & Mersey Canal*, and *Grand Trunk Canal*, were short titles. This company became the model for many later canal companies. As Charles Hadfield points out in *Canals of the West Midlands*, had the

carry coal all the way along the Bridgewater Canal to Manchester. From these boats Brindley developed the idea of using slightly larger vessels of comparable proportions for a long-distance canal. Initially he suggested boats 6 feet wide and 70 feet long which would carry 20 tons; then these dimensions were increased slightly and when the Trent & Mersey Canal was built, most of it was made to take boats 72 feet long and 7 feet wide. Many connecting canals were built to similar dimensions, and in this way evolved the narrow boat and the narrow canal, the two most distinctive features of the English canal system.

Boats of these dimensions could venture off canals and onto rivers, but were quite unseaworthy, which meant

that goods had to be transhipped wherever an estuary such as the Mersey formed part of a through route. But this disadvantage was outweighed by advantages. A narrow boat with a load of twenty five tons, hauled by a single horse, was large enough to show satisfactory economy compared with pack-horses or road wagons. The limited width of the canal meant economies in construction. Keeping the locks small meant low demand for water. Both the latter points were most important when long-distance, wholly-artificial canals were being built for the first time. The concept of the narrow canal with narrow boats seems to me to have been, for its era, both highly original (ships then were generally broad in proportion to their

The northern entrance to Blisworth Tunnel, Northamptonshire. Opening of this tunnel in 1805 marked completion of the main line of the Grand Junction Canal, and with it the direct canal line from the Midlands to London. The ground through which it was bored is wet and unstable and a first attempt to make the tunnel, started in 1793, had to be abandoned in 1796 because of flooding in the workings. The second, successful, attempt commenced in 1802. With a length of 3,056 yards it is the longest tunnel still in full use. It is a wide tunnel, but with no towpath.

One of the features of canals built late in the canal era is the aqueducts of cast iron which carry them over roads. This is Stretton Aqueduct, situated to the north of Wolverhampton and built in 1832, by which Telford's Birmingham & Liverpool Junction Canal crosses the Watling Street (grid reference SJ 873108). It is familiar to holidaymakers cruising on what is now known as the main line of the Shropshire Union Canal; hoardings which used to disfigure it have been removed in recent years, and nondescript grey paint has given way to the present attractive black and white.

Frindsbury Basin, Strood, in Kent, and the entrance lock of the Thames & Medway Canal; grid reference TQ 743695. Their alignment on Strood Tunnel, which a train is entering in the background, indicates the tunnel's canal origin. It was opened in 1824, two and a quarter miles long, as part of the canal, but in 1844 a single-track railway was built alongside the canal, and through the tunnel this was carried partly on the towpath and partly on a staging built out over the water. This curious example of dual use did not last long: within two years canal and railway had been bought by the South Eastern Railway Company which converted the tunnel entirely for the use of its North Kent Line, and reopened it as a railway tunnel in 1847.

The entrance into the great Tunnel, from Blisworth. Northamptonshire.

decision gone the other way, and organisation of canals by trusts similar to turnpike trusts become usual, then eventually waterways would, like highways, have been freed from tolls and have become the responsibility of national and local government. At the present time (and for the past fifty years) it would certainly be, and have been, very much in the public interest if canals were treated on an equal basis with roads. But once again one can scarcely blame the pioneers for lack of foresight to that extent.

The western goal at which both the Trent & Mersey and the Duke's canal were aiming was Liverpool. Wedgwood met John Gilbert who suggested to him that the T & M might link with the Duke's canal rather than the River Weaver, and this was agreed between the promoters of the T & M and the Duke. When the Trent & Mersey Canal obtained its Act in 1766, Parliament authorised a line from the Trent near Wilden Ferry to the Mersey estuary at Runcorn Gap: but the Duke was also authorised to alter his own intended line to Hempstones so that instead it would join the T & M at Preston Brook, from which place for six miles to Runcorn he would build the T & M line as part of his own canal at his own expense. In this way the Trent & Mersey was saved the cost of building part of its line, and got also convenient access to Manchester, as well as Liverpool via the estuary, and the Bridgewater Canal got a link to the South.

Earl Gower had agreed to head the Trent & Mersey company; the Duke of Bridgewater was a shareholder, and Samuel Egerton a substantial one. Josiah Wedgwood was appointed treasurer, and James Brindley surveyor-general; the clerk-of-works was Hugh Henshall, also an engineer, and to become a close associate of Brindley. With confidence that seems remarkable, but nevertheless proved eventually to be well placed, the builders of the Trent & Mersey started at the end of their line remote from the Bridgewater Canal. Derwent Mouth was the point selected for the canal to join the Trent; from here they worked westwards. The task was formidable: a canal $93\frac{1}{2}$ miles long with 109 road bridges, 164 aqueducts large and small, and five tunnels. The greatest and most remarkable of these was Harecastle Tunnel, 2,880 yards long

when completed, by which the canal pierced the ridge north-west of Stoke which forms the watershed between the east and west coasts of England. Construction of this proved extremely difficult and took nine years.

So long a tunnel was not strictly necessary for simple geographical reasons. This is confirmed by the subsequent history of tunnels in the vicinity. When a railway was built many years later it passed through a very long tunnel close to the original canal tunnel; but when this same railway was electrified during the 1960s, the engineers, rather than install overhead wires in a long and restricted tunnel, diverted the railway instead to the west, over a route which is nowhere more than half-a-mile from the original but which has only a comparatively short tunnel some 300 yards long.

There is no apparent reason why the canal should not have taken this course from the start, except that it would have meant more locks and a short summit pound. (A pound is a length of canal between two locks; a short pound at the summit of a canal, where it crosses a watershed, causes problems of water supply.) However, Harecastle Tunnel, as built, was analogous to the Worsley Mine tunnel on the Bridgewater Canal. It not only enabled the canal to pierce the ridge, but obtained for it a supply of water. Navigable soughs were driven from it to nearby coal-mines to drain them and provide traffic. The estate under which these mines were situated had been purchased as early as 1760 by a partnership which included John and Thomas Gilbert and James Brindley.

The Canals of the Cross

Long before the Trent & Mersey, or any part of it, was opened for traffic, the idea of a canal system started to spread further: Brindley's Grand Trunk Canal sprouted branches. Even before it obtained its Act, when it became clear that the branches considered earlier would not be included, other promoters planned a canal to link it to the Severn. This canal, the Staffordshire & Worcestershire, was authorised by Act of Parliament in

1766 on the same day as the Trent & Mersey: it was to run from Great Haywood on the T & M in Staffordshire for forty six miles to the Severn near Bewdley. James Brindley was its engineer, but much of the engineering detail was attended to by his assistants Samuel Simcock and Thomas Dadford.

A canal to link the town of Birmingham to the proposed Staffordshire & Worcestershire Canal was being considered as early as 1766, and in 1767 Brindley was asked to make a survey. His suggested route from Birmingham via Wolverhampton to join the S & W at Aldersley, with branches to Wednesbury and Ocker Hill, was adopted and an Act of Parliament obtained in 1768 for the Birmingham Canal. Construction started at once under, again, two of Brindley's assistants, Samuel Simcock and Robert Whitworth. The route of the canal, as planned and built, was very circuitous, winding about to follow the contours of the land. I discuss the contour canals more fully in chapter seven; this was the first of Brindley's contour canals to have an exaggeratedly tortuous course.

While the Birmingham Canal was of local interest, the Coventry Canal was of national concern. The idea of linking by water the rivers Trent, Mersey, Severn and Thames—and so the ports of Hull, Liverpool, Bristol and London—was not new, but it was taken up enthusiastically by Brindley who envisaged a canal network in the form of a cross to link the four rivers. The Trent & Mersey, which was to run in a south-westerly direction from the Trent to a point near Fradley, Staffordshire, and thence north-west to Preston Brook, provided the two upper arms, and the Staffordshire & Worcestershire one of the two lower arms. The fourth and final arm of the cross was to be provided, eventually, by the Coventry and Oxford Canals. A canal from the T & M to Coventry and Oxford was being considered in 1767 and early in 1768 the Coventry Canal's Act was passed by Parliament. The canal was to run from Fradley via Fazeley, Atherstone and Nuneaton to Coventry. James Brindley was once again engineer and surveyor, and construction started straight away. A little over a year later, in 1769, the Oxford Canal obtained its Act, to continue the line from a point near Coventry via Banbury to the Thames at Oxford. Brindley was engineer—he had surveyed the route in 1768—and Simcock his assistant.

Waterway developments elsewhere in Britain

By mid-1769, although all the canals just mentioned had been authorised and construction of many had started and was going on apace, no section of any canal of the cross had been opened to traffic. It is important, too, to remember that while the canal idea which germinated at Worsley was spreading so fast and far across the map of the Midlands, navigations elsewhere were still being built as a continuation of the steady build-up of waterway promotion which had been going on for the past 200 years. Only gradually did they come to owe more and more to Brindley and his ideas.

In 1760, for instance, an Act was passed to extend the navigation of the River Wey, which had been navigable from the Thames up to Guildford for over a century, upstream to Godalming. Smeaton was involved, and the four locks and four and a half miles of waterway were opened in 1763. In this year an Act was passed for the Louth Navigation. This had first been surveyed by John Grundy in 1756, to provide a link from Louth, Lincolnshire, to the sea and, via the Humber, the existing inland waterways of Yorkshire. His proposals were approved in 1760 by Smeaton who described the proposed navigation as a canal. As built, its first four miles with seven locks fell as a lateral canal beside the River Lud; then the canal deserted the river to run northwards across level country for seven miles so as to enter the sea at Tetney Haven. This was far enough up the coast for vessels from the Yorkshire inland navigations to reach it without leaving the Humber for the open sea. The lower part of the canal was opened in 1767.

The proposal for a canal to link the Firths of Forth and Clyde had been raised again at intervals since the time of Charles II, and in 1763 a Government-financed board asked Smeaton to make a further survey. He reported in 1764. That same year the Calder & Hebble Navigation was opened for sixteen miles from Wakefield up to Brighouse. But its finance was almost exhausted and what would today be called a boardroom battle ensued. Some of its commissioners wished to cease construction; others, based on Rochdale (who had already secured extension of the proposed line of navigation to Sowerby Bridge, whence Rochdale would be served by road) wished to continue. They won: in 1765 Smeaton was discharged, and replaced by Brindley, whose exploits at Barton would by then have been well known in nearby Rochdale.

Brindley proposed alterations to the line, which included construction of an aqueduct by which part of it, built as an artificial cut, was to be carried across the natural river. This proposal was not accepted, however, and it seems unlikely that Brindley, the protagonist of still-water canals, would have had the inclination to spend much time on a river navigation. When he was called in to advise commissioners appointed to make the River Soar navigable from the Trent up to Loughborough (they had obtained their Act of Parliament on the same busy day in 1766 when Parliament also authorised the Trent & Mersey and Staffordshire & Worcestershire Canals), he recommended, over much of the route, making a canal instead. Since the commissioners did not have powers to do this, the project rested.

Just when Brindley left the Calder & Hebble is not certain, but in 1766 he was also surveying for a canal to run from that waterway at Sowerby Bridge, across the Pennines to Rochdale, and on to join the Bridgewater Canal. As on the Soar though, there was no immediate result. In 1767 a succession of floods badly damaged the near-complete Calder & Hebble (a further section had been opened but had to be closed again); Smeaton returned to survey the damage and recommend improvements.

In the meantime a great dispute had been raging about the dimensions to which the proposed Forth & Clyde Canal should be built: the first (but not the last) occasion on which dimensions of a waterway were to become a matter of controversy. Smeaton had originally proposed a barge canal 27 miles long and 5 feet deep to enter the Clyde below Glasgow, near the highest point which ships could then reach. A rival group which included James Watt proposed a smaller canal, 20 miles long and only 4 feet deep, to enter the Clyde at Glasgow itself, while others again thought that a great canal, 7 to 10 feet deep, able to take seagoing vessels, would be best. The Act eventually passed in 1768 was for a canal 7 feet deep to run from the Forth near the mouth of the River Carron to the Clyde near Dalmuir, 35 miles, with a 3-mile branch to Glasgow. There were to be forty locks on the main line, each large enough to take vessels 66 feet long and 19 feet 8 inches wide. Even then opposition was not finished, and Brindley and others were retained to make a further survey of a small canal. This attempt was frustrated: construction of the great canal went ahead as authorised with Smeaton as engineer.

The careers of Smeaton and Brindley during the early stages of the canal era make an interesting contrast. Smeaton was the greater engineer; but he was already engaged on the Calder & Hebble at the time that the Duke of Bridgewater was promoting his canal—which must initially have seemed a comparatively small and insignificant undertaking. Though the Calder & Hebble was a river navigation, Smeaton soon afterwards undertook the Forth & Clyde Canal, which incorporated canal engineering works on the largest scale so far authorised in Britain. Yet Smeaton is remembered for his Eddystone Lighthouse, and Brindley's is the name associated with canals. He was the more ingenious of the two, the greater self-publicist, and the promoter of the concept of still-water canals as a widespread network.

How Brindley developed the idea of narrow boats and canals from the Worsley mine boats has already been described. But another line of canal development also originated from Worsley: the use of tub-boats, or very small craft, towed several at a time in trains as were the Worsley mine boats when hauled along the main canal to Manchester. Ducart's Canal in Ireland, on which work started in about 1765, was built in imitation of the Worsley system but was, as mentioned, a failure. A more successful imitator was Earl Gower who was of course

A marshy depression in the ground, bright with wild flowers, on the east side of the B2133, marks the course of the Wey & Arun Junction Canal 110 years after closure. The location is Loxwood, West Sussex, grid reference TQ 041311. The canal was opened in 1816 but, although it completed a through route between London and the south coast, was never very prosperous and was closed in 1868 a few years after a competing railway was opened. The towpath, on the right, has survived as a footpath.

3/5 (above) Squat and substantial, Great Haywood Aqueduct carries the Staffordshire & Worcestershire Canal over the Trent and is typical of Brindley's work.

most familiar with the Duke's canals. In the mid-1760s Earl Gower & Company (the company comprised Thomas and John Gilbert) built privately the Donnington Wood Canal in east Shropshire. It ran from a coal-mine on the Gower Estate at Donnington Wood near Oakengates for five and a half miles north-westwards to a roadside coal wharf near Newport, with short branches to limestone quarries and limeworks. The main line was completed by 1768. The tub boats used were 19 feet 8 inches long and 6 feet 4 inches wide: they carried 3 tons each. This canal became the first section of an intricate little system of tub-boat canals in the east Shropshire industrial area.

In Yorkshire work was proceeding on improving the navigation of the River Ouse above York, and in 1767 an Act of Parliament was passed to appoint commissioners to extend this route further by making the River Ure navigable upstream from the Ouse as far as Ripon. The top two miles were built as a lateral canal which became known as the Ripon Canal. The surveyor and engineer was a young man called William Jessop, later to become famous as a canal engineer but then aged about twenty-three. Jessop's father had worked under Smeaton on the Eddystone Lighthouse. After his early death, Smeaton became William Jessop's guardian, and took him on first as pupil, then as assistant.

Brindley, in 1768, was appointed Inspector of Works of the Droitwich Canal, which obtained its Act that year (he probably surveyed it during 1767). This was to be a 6¾-mile link from Droitwich to the River Severn; its locks were built to take river craft 64 feet long by 14 feet 6 inches beam. Later the same year Brindley was in Scotland surveying the abortive narrow rival to the Forth & Clyde Canal already mentioned, and then in the south of England to survey a canal from Salisbury to Redbridge on the Test estuary, which also was not built.

Brindley's canals are opened

In August 1769, however, came the first results of Brindley's work on the canals of the cross. It was three years since its first component, the Trent & Mersey, had been authorised, ten years since the Duke of Bridgewater's first Act of Parliament. But the section of canal which now came into use was far removed from these: it was six level miles of the Coventry Canal from Bedworth to Coventry city, on which construction had started just over a year before. The attraction of completing this section quickly was carriage of coal from mines to Coventry.

Any pride that Brindley may have felt at this achievement must have been tempered shortly afterwards, for the following month he was dismissed by the Coventry Canal Company. Excessive expenditure and, probably, too little of his time spent on the company's business were the reasons. Certainly, earlier that year he was, in addition to all his other commitments, surveying the Chesterfield Canal from Chesterfield via Worksop and East Retford to the Trent.

A greater achievement than that at Coventry came in November 1769 when ten miles of the Birmingham Canal were opened from Wednesbury to Birmingham. As at Coventry, and indeed Manchester, the immediate attraction to the promoters was to carry coal from mine to market, rather than the long-term formation of a canal network. This canal climbed over a summit at Smethwick, approached by flights of six locks on either side. Later alterations have eliminated nine of the total of twelve locks, but three of the original locks do still remain in use at Spon Lane: the oldest narrow locks. In terms of length of service, it is probable that the first narrow locks to be built were on the Staffs & Worcs, but this canal was not yet open.

A greater event still took place in June 1770 when the eastern part of the Trent & Mersey was opened, from Derwent Mouth to Shugborough. This meant thirty nine miles of canal opened at once, with twenty two locks, of which most were narrow but the lowest six were wide so that barges from the Trent might traverse the canal as far as Burton. A substantial nine-arched aqueduct carried the canal over the River Dove near that town, and there was a short tunnel through rock at Armitage. The aqueduct is still in use, but the tunnel, which must have been the first British canal tunnel brought into service apart from those giving access to mines, was opened out in 1971 as a consequence of mining subsidence. Its position (grid reference SK 071164) is marked by a short narrow steep-sided cutting, through which the canal is wide enough for a single boat only, spanned by a modern concrete road bridge which carries the A513.

The same year the Bridgewater Canal was extended for ten and a half miles to the outskirts of Stockton Heath (which was reached the following year). This section included a thirty-four-feet high embankment incorporating an aqueduct over the River Bollin, the highest embankment on the Bridgewater Canal. Just over 200 years later, on 2 August 1971, the embankment was breached, probably as a cumulative result of slight seepage through the canal bed over a long period. The escaping water washed a hole 90 feet wide in the embankment, and 20,000 cubic yards of material were deposited in the river bed. The canal was reinstated and re-opened in 1973: the site of the breach (grid reference SJ 728874) is indicated to passing boaters by the narrower-than-usual length of canal, edged with steel piles, which runs along the embankment.

The year 1770 was a year of openings, for as well as those mentioned, the Louth Canal was completed throughout in May, and the Calder & Hebble in September. It was also the year which saw the passing by Parliament of the Act for the Leeds & Liverpool Canal, which had been proposed in 1766 by John Longbotham, engineer, of Halifax, who had examined the Duke of Bridgewater's Canal. His survey had been checked and approved by Brindley and Robert Whitworth; Longbotham became engineer of the new canal. Parliament also passed the Act for the Monkland Canal, which was to carry coal into Glasgow from collieries to the east. In the South of England, Brindley was asked to advise on improvements to the lower Thames. Thinking bigger now than before, he proposed to bypass it with a canal large enough to take 200-ton barges, between Maidenhead and Isleworth. It was never built.

In March 1771 the short Droitwich Canal was opened, the first of Brindley's canals to be completed throughout since the Duke's canal had reached Manchester from Worsley six years earlier. The same month the first ten miles of the Oxford Canal were opened too (though as a result of an inter-company dispute they were not joined to the Coventry Canal for several years), and the Act of Parliament for the Chesterfield Canal was passed. John Grundy had surveyed a rival route to Brindley's which would

have been shorter and cheaper, but Brindley's route served more towns and was adopted. Construction work started, under Brindley's supervision, later in the year. The southern part of the Staffordshire & Worcestershire Canal, from Compton (near Wolverhampton) down to the Severn, was opened on 1 April; work on the rest was already far advanced.

In November 1771 the Trent & Mersey Canal reached Stone, and before the end of the year the Coventry Canal was open to Atherstone. There, for many years, it terminated: the company had exhausted its finance, and was already able to carry a profitable local coal traffic.

May 1772 saw the opening of the northern section of the Staffordshire & Worcestershire Canal, and this canal was then complete from the Severn to the Trent & Mersey which it joined at Great Haywood. Its line was just over forty six miles long, a narrow canal with forty three locks, four large aqueducts and many smaller ones, three short tunnels and a hundred or so overbridges. It had taken six years to build. On 21 November it gained an important branch when the Birmingham Canal was extended to join it at Aldersley by a line which served Wolverhampton and then fell by a flight of as many as twenty locks to its junction with the Staffs & Worcs. An additional lock was later added to complete the twenty one locks which remain in use.

By 1772 James Brindley had been overworking for years, often in arduous conditions. He was also suffering from diabetes, although this remained undiagnosed until, in September of that year, he caught a bad chill while surveying for the Caldon Branch of the Trent & Mersey. Within a few days he was seriously ill: on 27 September 1772 he died, aged 56.

In the short period of fourteen years he had laid the foundations of a national transport network. Much of it was unfinished at the time of his death, but three of his canals were complete (Droitwich, Staffs & Worcs and Birmingham) in addition to the original Worsley–Manchester section of the Bridgewater, and with the eastern part of the Trent & Mersey open also he had the satisfaction of knowing that two of the four arms of the cross were complete. The Trent was already linked by water to the Severn, the east coast of England to the west.

During the last years of narrow boat trade, a pair of empty boats heads north from Blisworth Tunnel, Northamptonshire, on the Grand Union Canal, a few hundred yards from the location of the illustration on page 26. In the background, a ledge high up on the right hand side of the cutting marks the course of the temporary tramroad that was completed in 1800, after the failure of the first attempt to make the tunnel, to link the two ends of the otherwise complete canal. The tramroad went out of use when the tunnel was opened, and the track was taken up and re-used on the line to Northampton.

Latter day boatwoman on the Grand Union Canal in the 1960s, surrounded by traditional narrow-boat decoration. Paintings of this type were already customary on boats in the 1850s: probably they derive from fashionable decorative motifs of the early nineteenth century, debased for popular use, adapted to canal boats and then forgotten elsewhere. Regular commercial traffic ceased on the Grand Union in 1970.

33

CHAPTER 4

THE YEARS OF EXPANSION

After Brindley

For the fifty years which followed Brindley's death in 1772, canals continued to multiply and continued to grow prosperous, with scarcely a hint of competition. So rapid and widespread did the increase in canals become that it is not simple to tell the story in an order which is roughly chronological without its appearing to become a jumbled list of facts with no geographical connection—but only by attempting to do so is it possible to demonstrate which canals were contemporaries in construction, which came early and which late.

Only a few days after Brindley's death the Trent & Mersey was opened from Stone up to Stoke—so completing the original scheme of 1758. Hugh Henshall took over as engineer to complete the canal. Another canal which (it was hoped) would join the Trent & Mersey at Middlewich had been authorised by Parliament earlier in the year. This was the Chester Canal, promoted by citizens of that port who were alarmed at the reduction of trade on the Dee which seemed likely to result from construction of canals to the Mersey. Their hope to see Chester connected to the Trent & Mersey Canal was, for a great many years, to be vain: even the 1772 Act insisted (because of T & M objections) that the two canals should not, at Middlewich, approach closer than 100 yards. But the connection may not have seemed so improbable at the time, for the Duke of Bridgewater was then having great difficulty in obtaining from an intransigent landowner the land needed to complete his canal. He did however start to construct the flight of wide locks which were to carry it down to the Mersey at Runcorn in 1772. They were completed at the turn of the year, and the canal was built inland from them, but by 1774 there was still a gap of a mile

past which land carriage had to be used. This situation was to continue for two years.

There was also controversy at this period over the Aire & Calder Navigation. This had altered little since built fifty years before, but had become very much busier. Excessive shallowness was the main complaint. Argument raged in and out of Parliament about what should be done: whether it should be improved, or whether a lateral canal should be built (two were independently proposed, one south of the river and one north). The eventual outcome was an Act of 1774 which, based on surveys by Smeaton and Jessop in 1772 and 1773, authorised several lengthy side canals on the upper part of the navigation and, on the lower part, an entirely new and wholly artificial route, the $5\frac{1}{4}$-mile Selby Canal, which gave a better outlet to the Ouse than the original. Construction of the Selby Canal started in 1775 with Jessop as engineer.

The Grand Canal

In 1772 also the Irish Parliament passed an Act which authorised incorporation of 'The Company of Undertakers of the Grand Canal'. Thomas Omer had built about ten miles of the canal between 1756 and 1763; then the Corporation of Dublin, with the concurrence of the Commissioners of Inland Navigation, had done further work on it, with a view to using it to supply Dublin with drinking water. But it was not progressing very fast towards the Shannon: the new company was authorised to acquire it and complete it, and to link it also with the River Barrow. A similar arrangement had already been made for the Limerick-to-Killaloe section of the Shannon Navigation, which was mostly lateral canal (though it later reverted to public ownership) and was

to be made in 1779 for the Lagan Navigation. The commissioners were disbanded in 1787, and most of their remaining waterways handed over to local commissioners: and subsequently both companies and government bodies have played a part in ownership of Irish waterways.

The new Grand Canal Company approached Smeaton and succeeded in persuading him to leave his work on the Forth & Clyde and Aire & Calder (he at first recommended John Grundy of the Louth Navigation in his place) and to travel to Ireland to inspect the Grand Canal works and line in 1773. He brought Jessop with him, and Jessop became consultant to the Grand Canal Co. for the next thirty years. Smeaton's report was helpful, and suggested building smaller locks than previously: he recommended a length of 60 feet and a width of 14 feet to take barges of dimensions similar to those in use on the busy and profitable Aire & Calder Navigation. In fact, locks were built slightly longer and narrower. Work on the canal restarted the same year under a succession of engineers the first of whom, John Trail, the new company inherited from Dublin Corporation.

Progress in Britain

The Forth & Clyde Canal was opened from the Forth as far as Kirkintilloch (about 20 miles and sixteen locks) by August 1773. The same month Smeaton resigned: for having brought the canal so far he received the grateful thanks of the company. The Ure Navigation, including the Ripon Canal, was opened the same year, and so was the first section of the Leeds & Liverpool, some thirteen miles from Bingley to Skipton east of the Pennines. The following year this section was extended eastwards to Shipley, including the five-lock staircase at Bingley, and westwards to Gargrave. Locks were built short but wide, 62 feet by 14 feet, to match vessels then in general use on Yorkshire waterways. There was progress also at the western end of the canal: it was opened from Liverpool to a junction with the Douglas Navigation, which the canal company now controlled, and which gave access to Wigan. The first section of the Chesterfield Canal too was opened in 1774.

In the south-west, work at last started on a project for which an Act of Parliament had been passed forty four years earlier, in 1730. This was the Stroudwater Navigation. The intention had been to make the Stroudwater river navigable from the Severn up to Stroud, but nothing further was done at the time, probably because of excessive opposition by mill owners. The proposal was revived in the 1750s, and in the early 1760s some work was done on a curious scheme to make the river navigable without locks, which removed the basis of the mill owners' objections: boats were to be restricted to their own reaches between mill dams, and cargo, carried in containers, transferred between them by crane. The work was abandoned before it was completed.

By 1774 enough canals had been built elsewhere for the Stroudwater proposal to be revived again, this time in the form of a lateral canal. It was thought this could be built under the 1730 Act for a river navigation which contained powers similar to those for the Sankey Brook. In this case, however, the opposition was strong and, after work had commenced, obtained an injunction to halt it. A new Act was successfully obtained the following year, which incorporated a company to build the navigation as, mainly, a canal wide enough to admit trows, the traditional river craft on the Severn. After this, work re-started.

Still further to the south-west, a canal had been proposed to carry sea sand from Bude, Cornwall, inland for use as fertiliser. An Act was obtained in 1774 but proved premature, and though the project was considered again in the 1790s, no construction was undertaken until much later.

The Bridgewater and the Trent & Mersey completed

There was progress in 1775 in the north-west. The Trent & Mersey was extended through Harecastle Tunnel to Sandbach and then Middlewich, and the Chester Canal opened a first section from Chester to Bunbury. The Trent & Mersey also completed its Preston Brook tunnel, and a short length of that canal's north-western extremity was opened, joining at Preston Brook the Bridgewater Canal which had reached that point, though it was still not open through to Runcorn. Soon afterwards, however, the Duke of Bridgewater was at last successful in obtaining the land he needed and, in the early part of 1776, work started on building the one mile of canal which was still needed to close the gap in the Runcorn line. This was completed, and the canal opened throughout, on 21 March 1776. It was seventeen years since the Duke had obtained his first Act of Parliament.

Another Act to make a navigation up the Soar from the Trent to Loughborough was passed in 1776. This authorised a river navigation with side cuts and a short canal for the final approach to Loughborough. The same year Acts were obtained for the Stourbridge Canal, to link the Staffs & Worcs Canal to that town and its district, and for the Dudley Canal to extend the line to Dudley. Both were narrow canals. An Act was passed, too, for the Caldon branch of the Trent & Mersey, to run from Etruria, Stoke, to Froghall in the Churnet Valley. Late in 1776, Huddersfield was linked to the Calder & Hebble by a three-mile long wide canal which had been authorised two years before.

4/1 The new Trent & Mersey Canal passes Wedgwood's Etruria Works, Stoke on Trent.

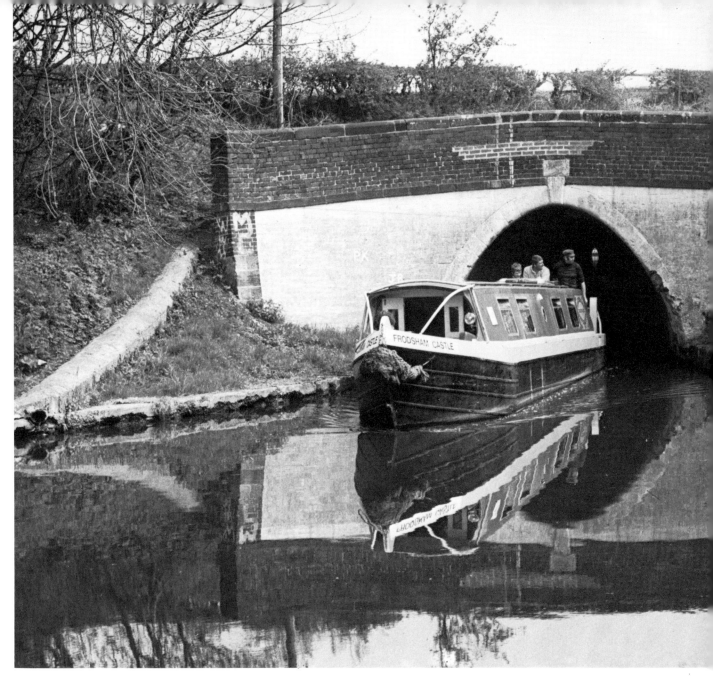

The Trent & Mersey Canal was completed throughout its main line in the spring of 1777, having taken eleven years to build. At both ends it was built as a wide canal to admit vessels from neighbouring waterways: flats from Preston Brook to Middlewich in the west and Trent barges as far as Burton in the east. The main, central, part of the canal was narrow. Later in the year another of Brindley's canals, the Chesterfield Canal, was completed too.

Construction of the eastern section of the Leeds & Liverpool had been continuing and in June 1777 it was ceremonially opened through to Leeds and a junction with the Aire & Calder. At the western end of the canal, work was concentrated on replacing the old Douglas Navigation by wholly artificial canal. At first, the canal from Liverpool had joined one of this river navigation's side canals at Newburgh, roughly half way along the Douglas

(which ran north-west from Wigan to the Ribble estuary). This side canal was incorporated into the Leeds & Liverpool line and extended to Wigan by 1780. In 1777 work started also on replacing the lower part of the Douglas by the Rufford branch of the canal. This was opened four years later.

Late in 1777 the Forth & Clyde Canal reached Glasgow from the Forth. The same year an Act was passed for the Erewash Canal, to run down the valley of the River Erewash on the Nottinghamshire–Derbyshire border, from Langley Mill to the Trent. It served collieries on the way and was laid out to enter the Trent opposite the mouth of the Soar so that coal barges from it might continue up the Soar to Loughborough. The Loughborough Navigation, and the first section of the Erewash, were opened about 1778. The locks on both were built large and long enough to take barges from the

Trent, or, in later years, narrow boats in pairs. In 1778 also the Oxford Canal reached Banbury on its way south, and the Selby Canal was completed to give the Aire & Calder its improved outlet to the Ouse. In this year also an Act of Parliament was passed for the Basingstoke Canal, to run from that town to the River Wey near Weybridge.

Tramroads

The Chesterfield Canal Company had bought a colliery in an attempt to build up traffic on its waterway and in 1778 constructed a wooden-railed tramroad from colliery to canal. A more ambitious venture of this nature came into use the following year, 1779 or, possibly, late in 1778: when the Caldon branch of the Trent & Mersey was

4/2 (left) The western portal of Barnton Tunnel, Trent & Mersey Canal (grid reference SJ 630748). There is no towpath, and boat horses would have been led over the hill. The hull shape of the emerging cruiser is based on the traditional narrow boat. Tunnels at the western end of the Trent & Mersey were built wide for Mersey flats but, probably, never used by them, for goods were transhipped between narrow boat and flat at Preston Brook.

4/3 (above) The southern section of the Oxford Canal is spanned by many wooden lifting bridges such as this. The bridge deck is balanced by timber beams.

opened to Froghall, its line was continued by a tramroad more than three miles long to Caldon Low itself. Carriage of limestone from Caldon Low quarries was the purpose of this canal-and-tramroad line, rugged country between Froghall and Caldon Low the reason for the change in the means of transport en route. Similar combinations of canal and tramroads were to be built subsequently in many locations during the canal era, for although canals were basically the more economic—a horse could haul a much greater load in a canal boat than in a tramroad wagon—tramroads were much cheaper to construct through hill country. The first tramroad from Froghall to Caldon Low had flat iron rails on top of wooden rails. It does not seem to have been satisfactory, for it had to be reconstructed more than once.

In 1779 also the Chester Canal was extended to Nantwich. It went no further. The capital was exhausted, as well as a large loan borrowed on the guarantee of Samuel Egerton, who was a shareholder and, it will be remembered, uncle of the Duke of Bridgewater. In this form as a local waterway with little trade the canal was hopelessly uneconomic: within a few years part of it was out of repair and disused. Much more successful were other canals completed the same year: the Stourbridge and Dudley Canals and the Erewash Canal.

Also in 1779 the first section of the Grand Canal in Ireland was at last opened to traffic, a length of nearly eighteen miles from Dublin to Sallins. It had taken a long time to build: but with fifteen locks, of which four were double locks, this was to be the most heavily locked section of the whole Grand Canal system as eventually built.

The line to Oxford

Though progress on the Grand Canal was slow, construction of the remaining incomplete arm of Brindley's Cross, the line from Fradley to Oxford, was slower still. Partly this was because of lack of money, partly because of lack of interest by the Coventry Canal. Its shareholders included local coal owners: the completed part of the canal carried their coal and was already profitable, to extend it would simply open their mines to competition. However in 1782 agreement was reached between the Coventry, Oxford and Trent & Mersey companies together with the promoters of a canal from Birmingham north-east to Fazeley (on the unbuilt part of the Coventry line). By this, the Oxford Canal Co. was to complete its line south to Oxford and the Coventry Canal Co. north as far as Fazeley. The remainder of the Coventry's line was to be built half by the Birmingham & Fazeley (as far as Whittington Brook) and half by the Trent & Mersey (onward to Fradley).

The Birmingham & Fazeley obtained its Act in 1783 and, though it had started out as a competitor of the Birmingham Canal, amalgamated with it in 1784. The combined company was at first called the Birmingham & Birmingham & Fazeley Canal Company, which cumbersome title was later altered to Birmingham Canal Navigations. Some branches of the Birmingham Canal had also been authorised and, with Smeaton as engineer, were built first; work on the Birmingham & Fazeley started about 1786, and re-started on the Oxford Canal in that year also. All these were narrow canals.

At this period four new waterways—Trent & Mersey, Erewash, Loughborough and Chesterfield—and the ancient Fossdyke all connected with the Trent, but little had been done to improve that river. In 1783 however an Act of Parliament established a company authorised to improve the Trent from Wilden Ferry down to Gainsborough, mainly by dredging, and to build a towpath. Jessop was engineer and the work was done by stages over the next few years.

The Thames & Severn

The idea of extending the Stroudwater Navigation by a canal over the Cotswolds to join the headwaters of the River Thames, and thus to link Thames with Severn, was probably as old as the idea of making the Stroudwater itself navigable. Two years after the Stroudwater had been opened, the company in 1781 had a survey made of such a canal, and then set about promoting it. The proposal attracted support from far afield—from, among others, London merchants and West Midland canal proprietors, the latter seeing the Thames-Severn link as encouraging increased traffic on their own undertakings.

The Act of Parliament for the Thames & Severn Canal was passed in 1783 and authorised a canal, to be built by a company distinct from the Stroudwater, to run for twenty eight miles from the end of the Stroudwater at Stroud, up the Golden Valley, and eventually to the Thames at Inglesham. There were to be forty four locks: for the first 2½ miles, as far as Brimscombe, the locks were to fit Severn trows, up to 15 feet 6 inches wide and 72 feet long, and over the rest of the route to take longer but narrower barges from the Thames, 86 feet long and 12 feet 2 inches wide. To pierce the watershed ridge at Sapperton, a very long tunnel was needed. Work seems to have started immediately; the engineer was Josiah Clowes, who had worked briefly on the Chester Canal, and the first section was opened in 1785, from Stroud up to Chalford.

At this time the Stourbridge and Dudley Canals were in effect a branch of the Staffordshire & Worcestershire Canal, but passage of the Birmingham & Fazeley and Thames & Severn Acts in 1783 prompted the idea that if the Dudley Canal could be extended to join the Birmingham, there would be additional traffic for both canals to and from both the north east and the south west. Between the Dudley and the Birmingham Canal lay high ground, but into this Lord Ward had already, in 1775, driven a private canal which branched from the Birmingham and led through a tunnel to his coal- and limestone mines. It terminated at Castle Mill Basin in a deep rock cutting. In 1785 the Dudley Canal Co. successfully obtained an Act to extend its line through a tunnel over one and a half miles long to Castle Mill. Here it was to join Lord Ward's canal, which was eventually incorporated into the Dudley Canal's line. Construction started at once, but it was to be several years before the tunnel was completed.

While these arterial canals were being built, another canal of local importance was authorised, one which, however, was to have its place in canal history. This was the Shropshire Canal, for which the Act of Parliament was passed in 1788. It was to run from a junction at the south end of the Donnington Wood Canal for nearly eight miles to a wharf beside the Severn at Coalport (not a junction: goods were transferred between boat

and trow). It was built as a tub-boat canal, and it overcame the problem of building canals through hilly country by using not locks but inclined planes, which transferred boats between its levels. A successful inclined plane had already been built on the Ketley Canal, a short private canal which became a branch of the Shropshire Canal, and the Shropshire Canal itself had three of them. Unlike those of Ducart's Canal, these inclined planes worked, and the canal was opened in sections over the next four years.

The years 1789 and 1790 were significant for canals, for they saw the completion of several important routes which had been under construction for many years. These in turn provided the impetus for promotion of many other canals, which culminated with the Canal Mania, a speculative bubble, in 1792.

The first of these routes to be opened was the Thames & Severn Canal, which took only six and a half years to build, a remarkably short space of time considering that the line included Sapperton Tunnel, 3,817 yards long, 15 feet wide and 15 feet high internally: the longest and largest canal tunnel so far built in Britain. With the opening of the Thames & Severn there was available for the first time a through, though roundabout, inland waterway route from the north and the Midlands to London. As a component of this, the canal was much hampered by the unimproved state of the rivers with which it connected. Lateral canals to bypass both the lower Severn and the upper Thames had indeed been proposed, but it was a great many years before they were built. The Thames proposal was successfully opposed by the Thames Commissioners, who did then make some improvements on the upper river itself, building pound locks to replace some of the flash locks. But not all of them, and in the 1790s traffic on the Thames was still regulated by a twice-weekly 'flash' of water—that is to say that, starting at Lechlade just below the junction with the Thames & Severn Canal, sufficient extra water was temporarily let past successive weirs to float barges over shallows in between. The flash travelled slowly downstream. It took about three days to reach Reading. In between flashes, when water was low, barges were likely to be grounded and their cargoes delayed.

The Cross completed

In any event the Thames & Severn's position as the sole through route from the North to London did not last long, for the fourth and final arm of Brindley's Cross was rapidly approaching completion. The Trent & Mersey had completed the Fradley Junction to Whittington Brook section in 1787 (it was then purchased by the Coventry company) and the Birmingham & Fazeley opened throughout from Birmingham via Fazeley to Whittington Brook in 1789. The Oxford Canal was completed from Banbury to Oxford on 1 January 1790 and the remaining gap was filled when the Coventry Canal was opened between Atherstone and Fazeley on 13 July. The canal cross was complete, twenty five years or so after Brindley had started to promote it.

Partly in anticipation of increased traffic from new connecting canals, and partly in response to traffic which was increasing anyway, the Birmingham company had decided in 1787 to imprive the line of the original Birmingham Canal at Smethwick. The short summit pound and flights of locks were proving to be a bottleneck: under Smeaton's direction a new summit pound was dug at a lower level than the original. It passed through what was for the period a deep cutting and enabled three of the locks on each side of the summit to be eliminated; it was completed in 1790.

In Scotland, work on the Forth & Clyde Canal had re-started about 1785, following an injection of Government finance and appointment of Robert Whitworth as engineer, and it too was opened through to the Clyde at Bowling in July 1790, completing the coast-to-coast link. A short extension in Glasgow, opened the following year, took it to its principal terminal there, called Port Dundas.

In the south of England, an Act had been passed in 1785 to extend the navigation of the River Arun upstream for thirteen miles to Newbridge: the work, mostly lateral canals (one of which included a tunnel) was completed in 1790. Two other widely separated canals were authorised in 1789. These were the Cromford Canal, to extend from the Erewash Canal up to Richard Arkwright's cotton-mill town of Cromford, Derbyshire, and the Royal Canal, to run from Dublin to the Shannon via Mullingar—a more northerly route than the Grand Canal, though one which had been considered and rejected when the Grand Canal was first planned and was now to be built to some extent in competition with it.

South Wales

At this point it is worth diverging farther than usual from chronological order, to consider the canals of South Wales and Monmouthshire. These, though they did not form a connected system, were closely concentrated in a small geographical area, and all the principal lines were built during the brief period of 1790 to 1799. There had been a few short canals built earlier than this in South-West-Wales, but the first canal of importance in South Wales was the Glamorganshire Canal, authorised in 1790 to link the ironworks of Merthyr Tydfil with the port of Cardiff. It was opened in 1794 and ran down the valley of the River Taf with fifty locks in its twenty five and a half miles.

Similar canals were then built in the valleys leading northwards from Neath and Swansea (the Neath and Swansea Canals respectively) and from Newport. The latter was the Monmouthshire Canal with two lines, one to Pontypool and the other to Crumlin. All these canals, totalling some seventy seven miles and 180 locks, were complete by 1799. Furthermore the Brecon & Abergavenny Canal had been authorised by Act of 1793 to extend the Monmouthshire Canal's Pontypool line for thirty-three miles to Brecon. This Act included powers to build tramroads up to eight miles long connecting with the canal (the Monmouthshire and Swansea Canal Acts had similar clauses) and in fact the B & A promoters built several tramroads first, before starting work on the canal proper in 1797. But then construction started at Gilwern, approximately mid-way along the line, and the canal was built northwards to distribute coal fed by the tramroads. It had reached Brecon before work started on closing the gap with the Monmouthshire Canal to the south.

Canal Mania

This activity in South Wales was matched on a far greater scale throughout much of England. From 1790 onwards, canal promotion boomed, reaching its peak in 1792 with the Canal Mania. When the dust started to settle about 1795, canals had been authorised to fill in the gaps in the canal map of England. They were in five fairly well defined groups, with several other important authorised canals distinct from these.

The largest group comprised the Grand Junction and Worcester & Birmingham Canals and the canals intended to connect with them or branch from them; and the greatest component of this was the Grand Junction. In 1791 London was still very much on the edge of the inland waterway map, dependent on the inadequate Thames. That year the Grand Junction Canal was proposed, to run from Brentford direct to the Oxford Canal at Braunston, a much shorter and more reliable route than the Thames. The Oxford Canal Company objected to this scheme, for it stood to lose traffic from the southern part of its route, and so proposed a rival line from Isleworth to Hampton Gay, only six miles north of Oxford. But the Grand Junction prevailed and, ninety three miles long, it was authorised by Act of parliament in 1793. James Barnes was resident engineer under William Jessop (whose mentor, Smeaton, had died the previous year). Two years later a branch to Paddington was authorised too.

In one particular the Grand Junction was a big step forward. Unlike earlier trunk canals, not only was it to be built as a wide canal throughout its main line, but its locks were to be long enough to take a pair of narrow boats. Alternatively they could take a 70-ton barge, 72 feet long by 14 feet 3 inches beam, suitable for the Thames. Or the Trent, or the Mersey also, for it was hoped that the wide canal would be extended to the Trent, and that existing narrow canals to the Mersey would be widened.

So far as a wide connection with the Trent was concerned, the Leicester Navigation, to make the Soar navigable from Loughborough onward up to Leicester, had already been authorised in 1791, and was being built.

Then, on the same day in 1793 as the Grand Junction, the Leicestershire & Northamptonshire Union Canal was authorised to build a wide canal onwards further from Leicester, to Market Harborough and Northampton, to which point the Grand Junction itself intended to build a wide branch canal.

In 1794 the Ashby Canal got its Act of Parliament. It was to run from a junction with the Coventry Canal at Marston near Bedworth to the coal-mining area round Ashby de la Zouch. It was hoped that it would be extended to the Trent, and it was hoped, too, that the Coventry and Oxford Canals would be widened between Marston and Braunston: so the Ashby Canal, lock-free, was built as a wide canal. What was not built, though it got as far as a Bill before Parliament, was the Commercial Canal, intended as a wide canal to run from the Ashby Canal to the Trent at Burton and on by Uttoxeter and the Potteries to join the Chester Canal at Nantwich. Since the Trent & Mersey had proved unwilling to widen, this would have provided a wide canal along the South-East to North-West axis of England.

The Worcester & Birmingham Canal got its Act of Parliament in 1791. It was to be a direct link between Birmingham and the Severn, and the promoters had ideas of building it wide enough for Severn trows. Construction started at Birmingham and wide barges and tunnels were built (there were three tunnels, all on the northern half of the canal) but economy eventually dictated narrow locks—fifty eight of them, all on the southern half. The canal was opposed by the Birmingham Canal which stood to lose traffic, but welcomed by the Dudley, which company evidently considered a canal outlet to Birmingham and the Severn independent of existing canals would be advantageous, and promoted a new line to join the W & B at Selly Oak, south of Birmingham. This was authorised in 1793.

In 1792 a canal was proposed which would run from the Worcester & Birmingham near Selly Oak to Stratford-upon-Avon with a branch to Warwick, so that coal from Dudley and merchandise from Birmingham could reach those towns. At this time the Grand Junction was also being promoted and the Birmingham canal company, seeing that a canal to War-

wick might be extended to join the Grand Junction, exerted its influence to get the Warwick & Birmingham Canal authorised as a distinct concern, connecting with the Birmingham & Fazeley Canal rather than the Worcester & Birmingham. Both the Warwick & Birmingham Canal and the Stratford Canal were then authorised by Parliament in 1793. The Stratford was to run from the Worcester & Birmingham at King's Norton (near Selly Oak) to Stratford-upon-Avon, and the two lines converged closely enough near Lapworth, Warwickshire, for a connection to be authorised. The Warwick & Birmingham was intended as a narrow canal, though it was later decided to build wide bridges and a wide tunnel. The Stratford Canal started off by building a tunnel wide enough for barges at King's Norton, but later built narrow locks further down the Canal. Clowes of the Thames & Severn was its first engineer.

4/4 John Rennie was engineer of the Lancaster and Kennet & Avon Canals, and endowed them with superb aqueducts over the Lune and the Bristol Avon.

A Warwick & Braunston Canal was authorised in 1794. Many of its promoters were identical to those of the Warwick & Birmingham. Then, to save construction cost, a further Act was obtained in 1796 to alter the line so as to run to a junction with the Oxford Canal at Napton, about six miles south of Braunston, and the name of the company was accordingly altered to

Warwick & Napton. To compensate the Oxford Canal for loss of traffic to the new route, the Oxford company was authorised to charge very high compensation tolls to traffic passing to and from the Warwick & Napton. The Warwick & Napton Canal was to be built as a narrow canal, for it rated narrow boat traffic from the Birmingham area more important than likely barge traffic to and from the Grand Junction.

The second group of canals with which promoters were busy during these years was in north-west England. In 1791 an Act was passed for the Manchester, Bolton & Bury Canal, to run from the Mersey & Irwell Navigation at Manchester north-west to Bolton with a branch to Bury. It was at first intended as a narrow canal, and construction started; then, as hopes grew of extending it to connect with the Leeds & Liverpool, and/or the proposed Lancaster and/or the proposed Rochdale Canals, the decision was taken to build wide, and narrow structures already built were widened.

Many years earlier Brindley had surveyed a canal from Manchester via Rochdale to the Calder & Hebble at Sowerby Bridge, and the project was revived as the Rochdale Canal in 1791. The line was re-surveyed by John Rennie, a Scots engineer who was already working on the proposed Kennet & Avon Canal mentioned below, but a Bill before Parliament in 1792 was lost because of objections by mill owners. Good transport was important to mill owners, but water supply was vital for it meant power, so that they often opposed canal projects which might divert their water. Just after the Rochdale Canal proposal was revived, another proposal was made for what at first appeared to be in effect a branch of it: the Ashton Canal, to run from the Rochdale Canal at Piccadilly, Manchester, to Ashton-under-Lyne with several subsidiary branches. It successfully obtained its Act in 1792, for a canal isolated, in the absence of the Rochdale Canal, from any others.

A long branch south-west from the Ashton Canal was also proposed, and was authorised separately in March 1794 as the Peak Forest Canal. It was to run from a junction at Dukinfield, near Ashton, via Marple to Chapel Milton, whence a tramroad would extend the line to limestone quarries that were proposed near Doveholes.

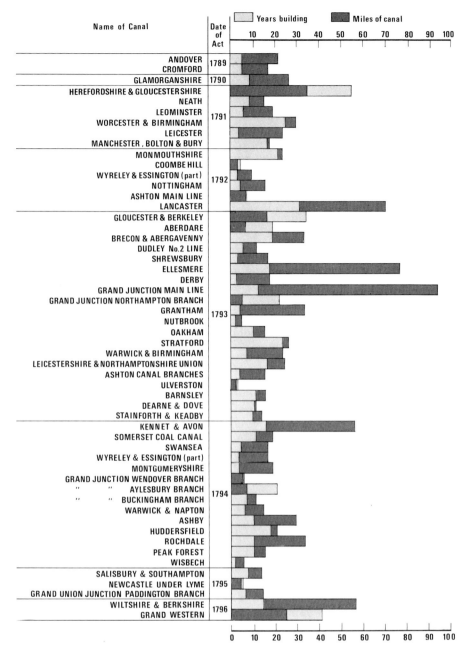

Name of Canal	Date of Act
ANDOVER	1789
CROMFORD	
GLAMORGANSHIRE	1790
HEREFORDSHIRE & GLOUCESTERSHIRE	
NEATH	
LEOMINSTER	1791
WORCESTER & BIRMINGHAM	
LEICESTER	
MANCHESTER, BOLTON & BURY	
MONMOUTHSHIRE	
COOMBE HILL	
WYRELEY & ESSINGTON (part)	1792
NOTTINGHAM	
ASHTON MAIN LINE	
LANCASTER	
GLOUCESTER & BERKELEY	
ABERDARE	
BRECON & ABERGAVENNY	
DUDLEY No.2 LINE	
SHREWSBURY	
ELLESMERE	
DERBY	
GRAND JUNCTION MAIN LINE	
GRAND JUNCTION NORTHAMPTON BRANCH	
GRANTHAM	1793
NUTBROOK	
OAKHAM	
STRATFORD	
WARWICK & BIRMINGHAM	
LEICESTERSHIRE & NORTHAMPTONSHIRE UNION	
ASHTON CANAL BRANCHES	
ULVERSTON	
BARNSLEY	
DEARNE & DOVE	
STAINFORTH & KEADBY	
KENNET & AVON	
SOMERSET COAL CANAL	
SWANSEA	
WYRELEY & ESSINGTON (part)	
MONTGOMERYSHIRE	
GRAND JUNCTION WENDOVER BRANCH	
″ ″ AYLESBURY BRANCH	1794
″ ″ BUCKINGHAM BRANCH	
WARWICK & NAPTON	
ASHBY	
HUDDERSFIELD	
ROCHDALE	
PEAK FOREST	
WISBECH	
SALISBURY & SOUTHAMPTON	
NEWCASTLE UNDER LYME	1795
GRAND UNION JUNCTION PADDINGTON BRANCH	
WILTSHIRE & BERKSHIRE	1796
GRAND WESTERN	

4/5 The Canal Mania: mileage of canals built in England and Wales as a result of Acts of Parliament passed during the years 1789 to 1796. Also listed are the canals which were built, either wholly or in part, together with the years in which their last sections were opened, and the mileage of each canal as built.

Speculative activity was at its height about 1792, to be translated into Acts of Parliament during the following two years. There were many other proposals for canals which were never built.

The canal/tramroad interchange point was later moved slightly nearer Manchester to Buxworth, and a short branch was added to Whaley Bridge. An even more ambitious extension of the Ashton Canal was the Huddersfield Canal from Ashton across the Pennines to Huddersfield, which town already had a wide canal down to the Calder & Hebble. The Huddersfield Canal was authorised on 4 April 1794: and the same day, such was the current enthusiasm for canals, Parliament passed an Act for another cross-Pennine canal—the Rochdale Canal at last, at its third application. The Ashton, Peak Forest and Huddersfield canals were all built as narrow canals, but the Rochdale Canal was built wide. Its locks took vessels 74 feet long

by 14 feet 2 inches wide; this was wasted effort so far as through traffic with the Calder & Hebble was concerned, for the latter's locks took craft only 57 feet 6 inches long by 14 feet wide. Jessop acted as consultant engineer during construction of the Rochdale Canal; Rennie by then was building the Lancaster Canal.

This had been authorised in 1792 to run from Kendal south by Lancaster, Preston and Chorley into the south Lancashire coalfield at Westhoughton, whence it was hoped to extend it to join either the Bridgewater Canal at Worsley or the MB & B Canal at Bolton, and so gain access to Manchester. This route brought the Lancaster Canal into conflict with the Leeds & Liverpool: the L & L had originally been authorised to run along the valley of the Ribble, but the company now wished to divert its line to run from its existing canal at Wigan northwards to Blackburn, and so over the Pennines as before. This proposed route was in parts very close to that of the Lancaster, and argument and dispute over which canal should be built where were to continue, in Parliament and out, for the next eighteen years. In 1795, however, the Duke of Bridgewater did get an Act of Parliament to extend his own canal westwards from Worsley to Leigh.

By 1792 the River Don in Yorkshire had been navigable from Tinsley downstream for forty years. Three proposals for canals based on it were made during the Mania: from Tinsley up to Sheffield, from Stainforth below Doncaster to Keadby on the Trent (the Stainforth & Keadby Canal), and from Swinton to Barnsley (the Dearne & Dove Canal). The latter was to be connected with the Barnsley Canal which in turn was to join the Calder & Hebble near Wakefield. Acts for all of these except the Sheffield proposals were passed in 1793.

Another product of the Canal Mania was the Ellesmere Canal, authorised by Act of Parliament in 1793. It was intended to link Mersey, Dee and Severn, and the line chosen (after various differences of opinion among the promoters) was from the Mersey estuary at the place subsequently to be called Ellesmere Port, across the Wirral to Chester as a wide canal to take Mersey flats, and onwards as a narrow canal through the broken country of the Welsh border, by Ruabon, with its coal-mines and iron works, and Chirk to the Severn at Shrewsbury. Branches were to serve, amongst other places, Llanymynech (limestone), Ellesmere itself, and Whitchurch. Around Ruabon, in particular, the engineering works were going to be big—a very long tunnel was intended, and the deep valleys of the Dee and the Ceiriog had to be crossed. So the company appointed William Jessop as Principal Engineer and then, under him to attend to detail, an up-and-coming architect-engineer named Thomas Telford.

Telford was a Scot, born in 1757, who served his apprenticeship as a stonemason at Langholm and then went to seek his fortune in London. There he worked on the construction of Somerset House under Robert Adam and his career evolved, through self-education and discerning patronage, into that of architect. In 1787 he was appointed Surveyor of Public Works for the County of Salop. The Ellesmere Canal was his first canal work, the first of much more, for as a canal builder he became famous, second only to Brindley; and as a pioneer civil engineer he came to rank second only to Smeaton, for as well as canals he was a great builder of roads and bridges.

Work on the Ellesmere Canal's Wirral section and the Llanymynech branch started in 1794; the same year an Act was passed for the Montgomeryshire Canal to extend the Llanymynech branch into Wales. In east Shropshire another canal, the Shrewsbury Canal, has been authorised in 1793. This was to be a tub-boat canal linking the Donnington Wood Canal with Shrewsbury to supply that town with coal. Clowes was engineer: then, in 1795, he died, and Telford was appointed in his place.

The other big trunk canal to come out of the Canal Mania was the Kennet & Avon. A canal link between the Kennet and the Bristol Avon (both of which had been made navigable in the 1720s) was being seriously considered in 1788 and the Act was passed in 1794. A wide canal engineered by John Rennie, it was to run for fifty seven miles from the Kennet at Newbury via Hungerford and Devizes to the Avon at Bath. The Somerset Coal Canal, planned to connect the K & A at Limpley Stoke with collieries south of Bath, was authorised on the same day.

Some attempts were made to improve the River Severn by dredging in 1791, and in 1793 an Act was passed for the Gloucester & Berkeley Canal. This was to be a lateral canal bypassing the river's winding and dangerous estuary and was to be built on the largest scale yet authorised in Britain, to allow small ships to reach Gloucester. The Crinan Canal also was authorised in 1793. This too was to be a ship canal, though not so large a one as the Gloucester & Berkeley, and by cutting across the Mull of Kintyre between Ardrishaig and Crinan would provide a short cut between the Clyde and the Scottish West Coast. Another proposal to cross a wider isthmus was the Grand Western Canal. It was first .considered in 1792 and obtained its Act in 1796, for a canal from the navigable River Tone at Taunton to Exeter, with a branch to Tiverton. One other canal of importance authorised during the Mania was the Nottingham Canal, to run from the Trent at Nottingham, up the valley of the River

Erewash (but on the opposite side to the Erewash Canal, with which it competed) to Langley Mill where it would join the Cromford Canal. Its Act of Parliament was passed in 1792.

Aftermath of a mania

All the canals just mentioned were built, either wholly or in part, so the Canal Mania, though largely speculative, did result in construction of much of the English canal system. There were of course many other proposals for canals, some of which were authorised but not built; while construction of those that were built often took far longer than anticipated. From 1792 onwards the British Isles were at war with France, and with the war came inflation. Estimates of the finance needed for entire canals proved adequate for a few miles only, and

4/6 (left) Tub boats were narrow and needed little headroom as is shown by this bridge (grid reference SJ 6721330) on the Shrewsbury Canal. The remains of a counterweight of a guillotine lock gate can be seen behind.

The year 1796 was a good one for openings. It saw also the opening of the Ashton Canal, as yet unconnected to the main system, the Peak Forest's tramroad and the top part of its canal, from Buxworth to Marple, the Nottingham Canal and the Beeston Cut. The latter formed part of the Trent Navigation and avoided a difficult stretch of river through Nottingham by diverging from the Nottingham Canal in the city and re-entering the river futher upstream. In Dublin, the Circular Line of the Grand Canal (in fact, semi-circular) was opened to link the Main Line of the canal to the tidal River Liffey at Ringsend where the Grand Canal Co. built docks for shipping. Parts of the Huddersfield Canal were opened during the period 1796 to 1799.

The Shrewsbury Canal was completed in 1797 (sections had been opened earlier) and part of the Montgomeryshire Canal was opened also. That year the Leicestershire & Northamptonshire Union reached Gumley Debdale in Leicestershire and there stuck, finance exhausted. The Trent & Mersey obtained Acts to build a short sub-branch from its Caldon branch to Leek with a large reservoir at Rudyard to supplement the water supply from Harecastle, and also to extend the Caldon branch to Uttoxeter. Part of the Ashby Canal was completed, and part of the Lancaster Canal, from Preston past Lancaster to Tewitfield. Early the following year a separate section of the Lancaster was opened from a point near Wigan as far as Chorley.

During 1798 the first sections were opened of the Kennet & Avon (Newbury to Hungerford) and the Rochdale Canal (Sowerby Bridge to Rochdale). The Dudley Canal's extension to join the Worcester & Birmingham was completed. By 1799 the Gloucester & Berkeley Canal had been completed for five and a half miles south of Gloucester and then, money already exhausted, construction ceased for several

there were long pauses in construction until more money was forthcoming. Work on the Grand Western Canal did not even start until 1810.

While the Canal Mania had been gathering momentum in England, construction of the Grand Canal in Ireland had been progressing well. It was opened in sections and, after work on the Shannon Line had been postponed in favour of that to the Barrow, the canal had been opened through to that river at Monasterevin in 1785. But navigation of the upper part of the Barrow was evidently unsatisfactory, for the canal was continued as a lateral canal to join the river further down at Athy, and was opened to that point in 1791. With this line of the canal complete, work went ahead on the Shannon Line. In the North of Ireland the Lagan Navigation was completed through to Lough Neagh in 1793; in Scotland the Monkland Canal was completed the same year, sections having been opened earlier. Its usefulness was giving rise to proposals for a

similar collieries-to-city canal to serve Edinburgh. These were to mature many years later as the Union Canal.

In England Dudley Tunnel had been opened to traffic in 1792. The year 1794 saw the opening of the Cromford Canal and completion of the Basingstoke, of which parts had already been opened. The Leicester Navigation was opened up the Soar from Loughborough to Leicester, and the following year the line was continued above Leicester with the opening of the first section of the Leicestershire & Northamptonshire Union Canal. Both the Manchester, Bolton & Bury Canal and the Wirral Line of the Ellesmere Canal were partly complete in 1795 and opened fully in 1796.

The Worcester & Birmingham Canal also was opened from Birmingham as fas as Selly Oak in 1795 and extended to King's Norton in 1796; the first and adjoining section of the Stratford was opened the same year. So was the northern section of the Grand Junction, from Braunston to Blisworth.

years. Construction of the Stratford, however, which had been at a standstill for three years, started again.

Tramroads and ship canals

Both the Warwick & Birmingham and the Warwick & Napton were completed in 1800. The Grand Junction Canal from the south had been opened in stages and in the same year reached the point which is now the foot of Stoke Bruerne locks. Between it and Blisworth was a gap of just over four miles, where a long tunnel was needed beneath Blisworth Hill. This was proving very difficult to construct and, as a temporary expedient, a tramroad was built to connect the two ends of the canal and brought into use in 1800. The Peak Forest Canal adopted a similar practice at the same time. Short of money to build locks—a single flight of sixteen locks near Marple was proposed—it built a temporary tramroad to bypass them and in this way opened its line throughout in 1800.

An Act of Parliament was passed in 1800 for the Thames & Medway Canal, which was to provide a short cut from the Thames at Gravesend to the Medway at Strood, both rivers being tidal at those points. Work started, but it was to go on in spasmodic bursts for many years before much progress was made.

In 1801 the Crinan Canal, having received a Government loan in 1799, was opened, though in an unfinished state. Further north, Telford was surveying for improvements to Highland communications by construction of roads, bridges and, he suggested, a ship canal to run through the Great Glen between Fort William and Inverness. A canal over the route had been considered before: the line included the large freshwater lakes of Loch Lochy, Loch Oich and Loch Ness, and artificial canal was needed only to link these to one another and to the sea. The Act for this canal, the Caledonian Canal, was passed in 1803. It did not incorporate a company but instead appointed commissioners responsible to Parliament to build the canal. Telford was appointed principal engineer with Jessop as consultant.

Meanwhile, in 1801, the Paddington branch of the Grand Junction Canal had been opened, from the main line at Bull's bridge, Southall, to Paddington Basin. On the other side of London, a curious thing was happening. There had been talk of canals there for some years, and in 1801 three Acts of Parliament were passed. One of these was orthodox enough: it was for the Grand Surrey Canal, which was intended to run from the Thames at Rotherhithe to Mitcham with various branches. Another route over which canal transport was needed was down the Wandle Valley for nine miles from Croydon to the Thames at Wandsworth. But Jessop, when consulted, pointed out to the promoters that a canal here was impracticable because it would require too much water from the River Wandle, leaving too little water to power its many mills. He recommended a tramroad instead, or as he called it, an 'iron railway': and so Parliament passed an Act not for a canal but for the Surrey Iron Railway. This was the first railway company, and Jessop was its engineer. Only a few weeks later, however, a further Act was passed which did authorise a canal for Croydon: the Croydon Canal, which was to run from that town to a junction with the Grand Surrey Canal at New Cross.

Two more tramroads built about this time were intended as temporary links past obstructions in lines of canal which were being built. The Lancaster Canal Company, faced with the need to cross the deep valley of the Ribble at Preston in order to connect the two parts of its canal, prepared plans for an aqueduct but instead built a tramroad, between Preston and Walton Summit, which the south end of the canal had reached. This tramroad was to become a permanent feature, for before the aqueduct could be built the railway era had overtaken that of canals. The Kennet & Avon built a tramroad, which did prove temporary, past the site of its great flight of locks at Devizes.

In 1802 the northern part of the Stratford Canal was completed, from its junction with the Worcester & Birmingham at King's Norton to its junction with the Warwick & Birmingham at Kingswood. The Leek branch of the Trent & Mersey was opened, and about this time the Stainforth & Keadby was opened too. This wide canal then became the principal outlet from the River Don to the tideway.

4/7 Date of construction was proudly carved on the keystone of the arch of a bridge on the Peak Forest Canal.

Death of a Duke

The Duke of Bridgewater had seen his canal complete and prospering, and the canal idea spreading far and wide from it. Then, in 1803, the grand old man of canals, he died. The dukedom died with him. In death as in life he left his mark: ownership of the Bridgewater Canal was vested in a complex trust for the benefit of his heir, but a manager of the canal was appointed with power to appoint his own successor. The same year the Surrey Iron Railway was opened.

4/8 The Ellesmere Canal's Llangollen branch at grid reference SJ 241423. The small arch in the bridge, to the right of the canal, formerly accommodated a tramroad.

An Act was obtained in 1803 for the Tavistock Canal, to connect Tavistock, and nearby copper-mines and slate quarries, with Morewellham on the Tamar estuary. The main line of this tub-boat canal, though only four miles long, was to include a tunnel of nearly one and a half miles. Consequently the canal took many years to construct.

On the Peak Forest Canal, Marple Locks were completed in 1804 and the temporary tramroad went out of use. The same year the Rochdale Canal was opened throughout: the first trans-Pennine canal to be completed, though it had been the last to be authorised.

The Ashby Canal was opened in 1804, having earlier built tramroads in place of intended branches. The effort of building it as a wide canal proved to be largely in vain, however, for wide boats could be used only internally. The Coventry and Oxford Canals to Braunston were never widened, and although Blisworth Tunnel was completed in 1805, and with it the main line of the Grand Junction Canal, that company soon found that the hindrance to traffic caused by wide boats passing through the long tunnels at Blisworth and Braunston was so great that, in the absence of a wide canal continuing from Braunston, it shortly afterwards prohibited wide boats from using either tunnel. The tunnels were wide enough for narrow boats to pass one another. The Blisworth Hill tramroad went out of use and the company thriftily re-used the materials to build a tramroad from the canal to Northampton in place of the proposed branch.

Another trunk route was completed in 1805: the Grand Canal in Ireland was opened through to the River Shannon. After the line to the Barrow had been completed, construction of the Shannon Line had gone ahead, aided by Government grants. The junction with the existing canal was at Lowtown, near Robertstown, Co. Kildare, and the canal was built and opened in stages via Tullamore to Shannon Harbour, north of Banagher.

Towards the end of 1805 another important link was made: the Ellesmere Canal was joined to the Chester Canal though not, as was originally intended, at Chester, but at Hurleston near Nantwich. Plans had changed. Apart from the Wirral Line, only seventeen and a half miles of the intended main line were built, from Trevor southwards to Weston Lullingfield. The completion of the Shrewsbury Canal, which supplied Shrewsbury with coal from East Shropshire, had made it no longer worthwhile to complete the Ellesmere Canal to that town; similarly, north of Trevor, the ruggedness of the country, and changes in the pattern of coal-mining no longer made it worthwhile to complete the line to Chester. Part of it immediately north of Trevor was built as a tramroad and a branch from Trevor to the Dee above Llangollen drew water from the river to feed the canal. To connect the two parts of its canal, the company had chosen the cheaper option of building its branch eastwards to Ellesmere and Whitchurch and extending this to join the Chester Canal near Nantwich. Even so, the combined system of the Ellesmere and Chester Canals was isolated from the rest of the canal network. The section of main line that was built had provided canals with their greatest engineering feature, which impresses today as much as ever: the Pontcysyllte Aqueduct, 1,007 feet long, which carries the canal 127 feet above the River Dee. Of this, more in chapters seven and ten.

The Royal Military Canal in Kent was built during the period 1804 to 1806 primarily as a defensive work. There was much concern then about the possibility of invasion by the French, and Romney Marsh was thought the most probable location. The canal, therefore, was built inland between Hythe and the River Rother near Rye to present a barrier to an invading army. It was never needed as such, though it did have some usefulness for local transport. In 1806 too an Act of Parliament was passed for the Glasgow, Paisley & Ardrossan Canal.

The Grand Surrey Canal Company, after spending some years considering whether or not to become a dock

undertaking instead, opened its canal from the Thames as far as the junction with the Croydon Canal and on to the Old Kent Road in 1807. The Croydon Canal itself was opened in 1809.

Completions and expansions

In the East Midlands, at this time, there was still a large gap between the Grand Junction and the Leicestershire & Northamptonshire Union Canals. The latter canal in 1809 opened an extension as far as Market Harborough, and various possibilities for completing the line were considered. The eventual choice, authorised by Parliament in 1810, was the Grand Union Canal, to run from the Grand Junction at Norton to the L & NU at Foxton. Bridges and tunnels were to be built wide but locks were to be narrow, because of the prohibition of wide boats on the northern part of the Grand Junction.

The same year the Kennet & Avon Canal was open throughout. The canal company had already obtained control of the Bristol Avon, and later acquired the Kennet, after which the whole waterway from Reading to Bristol came to be called the Kennet & Avon

Canal. The eastern part of the Leeds & Liverpool was extended in 1810 as far as Blackburn, and the L & L company then settled its differences with the Lancaster Canal and agreed to complete its own route by joining the existing south end of the Lancaster Canal at Johnson's Hillock (near Walton Summit) and diverging from it again ten miles further south near Wigan, to connect with its own western section.

This was a period when people once again began to consider far-reaching extensions to the canal system. There were proposals, among others, for canals from Market Harborough to Stamford, to connect with the antique Stamford Canal and the River Welland; from Paddington to the Thames tideway; and (after years of talk) to link the Wey and the Arun Navigations and so provide a route from London to the south coast. The first of these did not get through Parliament, but the latter two did, as the Regent's Canal, authorised in 1812, and the Wey & Arun Junction Canal, authorised in 1813.

While this was going on, other canals were being completed. The Huddersfield Canal was opened in 1811: Standedge Tunnel, through the Pennine ridge, had taken seventeen years to construct and was, at 5,456

4/9 (above) A contemporary view of the Croydon Canal in the rural surroundings of Forest Hill. The canal was converted into a railway in the late 1830s.

4/10 (right) Despite the upheaval that they must have caused, contemporary views of canals being built are extremely rare. This one shows Islington Tunnel on the Regent's Canal under construction.

4/11 (overleaf) Early narrow boats are clearly shown in this engraving of the junction of the Regent's Canal (left background) with the Grand Junction Canal at the place now known as Little Venice. Crews appear to be all-male.

yards, the longest canal tunnel ever built in Britain. The Trent & Mersey branch to Uttoxeter was opened the same year, and the Glasgow, Paisley & Ardrossan Canal as far as Johnstone, the farthest it ever reached. This little canal was isolated from all others, but was later to make its mark in canal history with the development of swift passenger boats.

The Brecon & Abergavenny Canal had opened its northern part many years earlier and supplied Brecon with coal originating on the associated tramroads: it then built its southern section and eventually was open throughout to the junction with the Monmouthshire Canal at Pontymoile, near Pontypool, in 1812. Throughout the hills and valleys of South Wales and the Border Counties, the mileage of tramroads built was eventually far greater than that of the canals with which they were connected. They extended as far north as Hereford and Kington.

In 1812 construction of the Stratford Canal again re-started, after an interval of ten years. The following year the Ellesmere and Chester Canal Companies amalgamated.

The Grand Western Canal was opened between Tiverton and Holcombe Rogus in 1814: this isolated section, intended as a branch and a short piece of the Taunton–Exeter main line, had been built first to carry limestone quarried near Holcombe. It was built on a grand scale, however, with a clear width of water beneath bridges of eighteen feet.

The Grand Union Canal from Norton to Foxton was opened in 1814, completing the route from the Grand Junction to the Trent, and the Grand Junction's Aylesbury branch, authorised as long ago as 1793, was opened in 1815, as a narrow canal. The Grand Junction also opened a narrow branch to Northampton, replacing the tramroad. At Northampton it entered the navigable River Nene. The same year the Worcester & Birmingham was at last open throughout from Birmingham to the River Severn.

Two projects long considered in Yorkshire came nearer fruition in 1815 when Acts were passed for canals from the River Don up to Sheffield, and from the River Derwent to Pocklington. Another Act authorised the extension of the Montgomeryshire Canal to Newtown by a separate company.

In 1816, twenty three years after its Act of Parliament, the Stratford Canal was completed through to Stratford and the River Avon. The Leeds & Liverpool had been even longer building, but that year the connections with the Lancaster Canal were opened and the line from Leeds to Liverpool was complete. The Wey & Arun Junction Canal also was opened in 1816, eighteen and a half miles long from Shalford on the Wey to Newbridge at the head of the Arun Navigation, with twenty-three locks able to take boats 74 feet 9 inches long and 13 feet beam. Part of the Regent's Canal too was opened in 1816.

The Tavistock Canal was opened in 1817, and the Royal Canal was completed through to the Shannon Navigation, the second route from Dublin to that river. Acts of Parliament were passed that year for the Portsmouth &

4/12 (right) Dublin seen from Blaquiere Bridge on the Royal Canal branch to Broadstone Harbour, most of which has long since been filled in.

Arundel Canal which was intended to continue the Wey & Arun line to Portsmouth, and for the Edinburgh & Glasgow Union Canal (or Union Canal for short) to link the Forth & Clyde Canal near Falkirk with Edinburgh. The Poor Employment Act of 1817 enabled Government money to be lent for public works to relieve unemployment, and loans made under this Act enabled work to re-start on the Regent's Canal (opened in 1820) and on the Gloucester & Berkeley—rare instances in which Government money was used to build canals in England.

The Pocklington Canal was opened in 1818 (an equally rare instance of a canal built for less than the estimated cost). The eastern section of the Caledonian Canal was opened also, from the sea to Loch Ness, and by continuing up the lock ships could reach Fort Augustus at its head.

The Sheffield Canal was opened in 1819. Interest had revived a year or two earlier in the Bude Canal and an Act of Parliament for the Bude Harbour and Canal Company was passed in 1819 also. Another canal authorised the same year was a branch of the Leeds & Liverpool to connect Wigan with the Bridgewater Canal at Leigh; it

was opened in 1821, filling in a conspicuous gap in the waterway system of the north west. But its construction meant too that not only would the final southern extension of the Lancaster Canal not be built, but also that the Manchester, Bolton & Bury was frustrated of all its hopes for direct connection with other canals.

In 1821 the Montgomeryshire Canal to Newtown was completed and the Aire & Calder Navigation, needing a more direct outlet to the Humber than that by Selby and a more reliable one than the lower River Aire, obtained an Act for a lateral canal from the Aire at Knottingley to the Humber at Goole.

The following year the Caledonian Canal was opened throughout, though not to the full dimensions originally envisaged, and it had later to be closed for some years and rebuilt. In 1822 also the Union Canal was opened and Edinburgh was linked to Glasgow by canal. The Union Canal's aqueducts rank for size and grandeur second only to Pontycysyllte.

The Bude Canal was opened in 1823. It commenced at Bude with a sea lock able to pass coasting vessels, above which were a couple of miles of barge canal. But the rest of the canal was tub-boat canal, thirty three and a half miles long with three branches and six inclined planes: the most extensive tub-boat canal ever built.

Harecastle Tunnel on the Trent & Mersey, once the wonder of the waterways, had over the years become a hopeless bottleneck. John Rennie had recommended to the Trent & Mersey Company in 1820 that it should be duplicated, and, after his death, Telford was called in for another survey. He advised that a new tunnel, of larger bore than the original and with a towpath, should be made parallel to the existing tunnel. An Act of Parliament was obtained for this in 1823 and Telford was placed in overall charge of construction with a resident engineer responsible to him.

At this point, with canals still prosperous and expanding, I will pause in this headlong gallop through canal history, for much was soon to change. I have endeavoured to show how fast and far and extensively canals spread, though even so many canals of mainly local significance have had to be omitted: let us next see what traffic the canals were carrying, and how they organised it.

CHAPTER 5

TRAFFIC, RAILWAYS AND LATE CANALS

Canal traffic

'It is impossible to describe to you how flourishing the canal is,' wrote Thomas Dundas to his father Sir Lawrence in 1777, 'there are more ships at the sea lock than at Bo'ness and Alloa put together.' The canal was the Forth & Clyde, at that date incomplete, but open from the Forth to the outskirts of Glasgow. Sir Lawrence Dundas was one of the principal promoters of the canal, his son eventually governor of the company for thirty years. The particular trade which was then helping the canal to flourish was timber, imported from the Baltic and dispatched along the canal for use in Glasgow.

Timber was one of the staples of canal traffic, but most important of all was coal. If a canal was to prosper, said the Duke of Bridgewater, it needed 'coals at the heels of it'. Other traffics might fluctuate, but coal remained steady. There can have been scarcely a canal where coal traffic was not vital. The canals which were best off were those which not only served coal-mines but were able also to distribute coal to towns and cities along their own routes. They included the Bridgewater and also the Birmingham, the Coventry, the Trent & Mersey, the Ashton and others. Some canals such as the Ashby, Erewash and Somerset Coal Canals served coal-mines but, having no large towns on their own routes, sent their coal farther afield. Elsewhere places remote from mines received coal from their local canals—places such as Oxford, Stratford, Croydon and Lancaster, and the towns along the Grand Junction and Kennet & Avon Canals.

After coal, probably the most important bulk traffic was limestone. It was used as a flux in smelting iron, and burned in kilns (often located beside canals) to produce lime for use as a fertiliser and as a constituent of mortar. Limestone traffic was behind construction of the Peak Forest Canal, and the Caldon branch of the Trent & Mersey; it was important also on the Ellesmere, Cromford, Dudley, Stratford and Leeds & Liverpool Canals among others.

In agricultural areas the produce of the country, particularly grain, provided much traffic. Corn from Lincolnshire, for instance, was carried over the Fossdyke, the Trent, the Aire & Calder and Calder & Hebble Navigations and the Rochdale Canal to feed industrial Lancashire. Corn and grain were also important traffics on the Grand Junction, Stratford, Trent & Mersey and Forth & Clyde Canals. Other important bulk cargoes were hay and building materials. General merchandise, parcels and groceries were carried almost everywhere.

Of comparatively specialised traffics, iron was important on those canals which served ironworks, particularly the South Wales canals, the Cromford and the Forth & Clyde. Clay, pottery and glass were much carried on the Trent & Mersey and Staffs & Worcs Canals. Relatively unusual cargoes were hops on the Thames & Medway (opened in 1824), and slates and herrings on the Crinan. The Grand Canal in Ireland carried all the traffic of an agricultural country, though peat was carried in much larger amounts than coal, and Guinness's porter became increasingly important.

Canals and carriers

The Duke of Bridgewater carried coal and other goods in his own boats, and the Birmingham Canal seems to have had the coal trade nearly sewn up in the early days, using its own boats too. But generally canal companies provided the highway, and carriers provided the boats and did the carrying, as had been the practice on earlier river navigations. Into canal companies' Acts of Incorporation was written a public right of navigation, that their canals should be available to all who wished to use them, on payment of

5/1 William Brocas sketched the Grand Canal in Dublin from life, with Portobello canal hotel and a passage boat preparing to leave, in watercolours too subtle to be reproduced. This drawing is copied from a photograph of the original. See also colour plate on page 178.

appropriate tolls. It was a *quid pro quo* for the powers of compulsory purchase of land which they were also granted. Many canal companies were not permitted to do their own carrying, lest they show undue preference for their own boats at the expense of other carriers.

Of carrying fleets from the early days, the most familiar name is that of Pickfords. They were road carriers even before they started to carry by water; then they built up an extensive fleet using not only narrow boats but also Severn trows. Some canal builders set up in business as carriers: Hugh Henshall of the Trent & Mersey was one, though his firm, Hugh Henshall & Co., later became a subsidiary of the Trent & Mersey Canal Co. John Gilbert went into partnership with a road carrier to set up the canal carrying firm of Worthington & Gilbert.

Bulk cargoes were carried in heavy-laden, slow-moving craft, but merchandise and parcels were carried by fly boats, lightly loaded vessels hauled day and night by relays of horses and operating to something resembling a timetable.

Passenger traffic

Passengers had travelled by river for centuries, and passenger boats were introduced on the new river navigation of the eighteenth century. They were running on the Bristol Avon, for instance, by the 1720s. Passenger carrying on canals started in 1767 when converted barges were put into service on the Bridgewater Canal between Altrincham and Manchester. They were very successful, and the service was later extended, with boats built for the purpose, to cover the whole route from Runcorn to Manchester, and from Worsley to Manchester also. Similar passenger services followed on many canals in north-west England: the Chester Canal, the Wirral Line of the Ellesmere Canal, the Manchester, Bolton & Bury, the Ashton, the Leeds & Liverpool and the Lancaster.

Canal boat travel must have offered a pleasant jolt- and dust-free alternative to road travel, whether by carriage or on horseback. It is surprising that it did not become more widespread in England apart from the north-west, although there were passenger boats between Paddington and Uxbridge on the Grand Junction, between Bath and Bradford-on-Avon on the Kennet & Avon, and between Nottingham and Cromford. Great improvements in both coaching and road surfaces were being made in England during the period that the canal system was evolving, and these may have had something to do with it. Certainly, apart from carriage of soldiers between London and Liverpool, there was nothing in England to compare with the long-distance passenger services operated by the Grand Canal in Ireland.

The first passenger boats, or passage boats as they were called locally, on the Grand Canal entered service in 1780 within a few months of the opening of the first section of the canal to traffic. The service was successful, further boats were obtained, and as the canal was extended, so were the passage boat services. In due course they ran between Dublin and Athy, and Dublin and Shannon Harbour, and when a branch line was opened beyond the Shannon to Ballinasloe the passage boats went there too. Road coaches connected with the passage boats at Tullamore, Shannon Harbour and Ballinasloe. The Royal Canal had long-distance passenger services also, between Dublin, Mullingar and Longford, which was reached by a short branch from the main line, and there was a passenger service on the Newry Canal between Newry and Knock Bridge near Portadown.

In Scotland also passenger services were widespread on canals. The Forth & Clyde had 'track boats' carrying passengers and goods by the late 1780s if not earlier. The passenger service ran from Glasgow to Lock 16 near Falkirk. This was the top of the heavily locked section leading down to the Forth; at Lock 16, the track boats connected with coaches to and from Edinburgh. In 1809 purely passenger boats began to operate, and when the Union Canal was opened, through passenger services started between Glasgow and Edinburgh. There was an overnight boat as well as the daytime ones.

The Glasgow, Paisley & Ardrossan and Monkland Canals also carried passengers. Experiments on the Paisley Canal in the early 1830s resulted in development of long, light passenger boats able to travel at speeds above ten miles per hour. An intensive passenger service was operated, which competed effectively with road coaches. 'Swift' passenger boats of the same type were subsequently introduced on the Forth & Clyde and Union Canals in Scotland, the Lancaster and Kennet & Avon Canals in England, and the Grand and Royal Canals in Ireland, in each case with beneficial effect on journey times. In Ireland such boats were called fly boats, the fast merchandise-carrying boats called fly boats in England being unknown.

The schedules were indeed, for boats drawn by horses, astonishing: Dublin to Tullamore, for instance, fifty-eight miles and twenty six locks, in nine hours and five minutes; or Preston to Lancaster, thirty lock-free miles, in three hours and twenty minutes. A moment's calculation shows that to travel thirty miles today on a British Waterways Board canal, with its four miles per hour speed limit, would take at least seven and a half hours. In practice it would take a couple of hours longer still.

Passenger traffic on the Crinan and Caledonian Canals was of a different nature. The first steamboat to enter commercial service in Europe, Henry Bell's *Comet*, started to ply on the Firth of Clyde in 1812, and about 1819 she commenced running between Glasgow and Fort William, passing through the Crinan Canal on the way. She was the first of many steamers to operate this route. The following year Bell started a steamer service on the eastern end of the Caledonian Canal, from Inverness to Fort Augustus, and because the canal was not yet open throughout, there was a road connection from Fort Augustus to Fort William, where passengers could catch the steamer to Glasgow. As soon as the remainder of the Caledonian Canal was opened in 1822 steamers started to operate between Glasgow, Oban, Fort William and Inverness, passing through the two canals en route. There was an element of paradox in this, because steam power for ships, which developed during the period that the Caledonian Canal was being built, was soon to remove one of the main reasons for its construction—to enable sailing ships to avoid the dangerous and difficult passage round the north of Scotland.

As of most other things about

5/2 *Thomas Telford.*

5/3 *Among Telford's greatest works was the new main line of the Birmingham Canal Navigations, seen here spanned by his cast iron Galton Bridge. This still survives (grid reference SP 015894).*

canals, the Duke of Bridgewater can be considered the father of pleasure boating. He not only had his own 'gondola', used for pleasure and entertainment of important visitors such as Josiah Wedgwood, but encouraged friends to put pleasure craft on his canal too. It was not uncommon for landowners to insist on the right to have a pleasure boat, free of toll, on a canal made through their property. And though the real era of canals-for-pleasure was to come much later, rowing boats could at this period be hired on the Croydon Canal, which contemporary prints depict passing through pleasant countryside near London and on the Thames & Medway Canal; excursions by steamboat operated on this canal, and by horse-drawn passenger boat on the Union Canal at Edinburgh.

Origins of steam railways

The early 1820s saw the canal network largely complete, with canals in general very busy and prosperous and planning improvements either willingly or of necessity. But while the canal system had been developing, use of steam for power had been developing too. Steam power had long been used for pumps: canals were great users of steam pumps for water supply. From these, beam engines had been developed to power stationary machinery. Early steam pumps and engines were bulky and clumsy: but steam engines had recently been refined sufficiently to power ships, as just mentioned. The continuing process of

development was about to produce steam plant compact enough and reliable enough to power land vehicles, on railways and also, it seemed possible, on ordinary roads. Already primitive steam locomotives were at work on tramroads. Some of these were canal feeders, others served collieries in the north-east of England.

One of the first people to see which way the wind was blowing was William James. James had been behind resumption of work on the Stratford Canal in 1812, which resulted in the southern part of the canal being completed through to Stratford-upon-Avon. He was already owner of the Upper Avon Navigation, from Stratford to Evesham. But he was also interested in tramroads, on the grand scale. Stratford was to become, according to his plans, an important

interchange point: the line of the Stratford Canal would be continued by the 'Central Junction Railway or Tram-road' to Moreton-in-Marsh, Oxford, Uxbridge and London; a branch, from collieries which James owned in Warwickshire, would join the main line near Shipston-on-Stour.

This scheme was too grandiose, too premature, and seems today too round-about a route from the Midlands to London. It is worth noting, however, that it was indeed sound in one respect. Across the route of any line from the Midlands to London lay the range of hills which starts as the Cotswolds and continues north-east through Northamptonshire: canals already built had penetrated this by long and costly tunnels at Sapperton, Fenny Compton and Braunston, and the eventual London & Birmingham Rail-

way was to do likewise at Kilsby. Alone among the promoters of these lines, James identified a route which needed no such tunnel.

Though the Central Junction was too ambitious, a local tramroad, to extend the line of the canal to Moreton and Shipston, was not. It was incorporated by Act of Parliament in 1821 as the Stratford & Moreton Railway Company. James had been in touch with George Stephenson, whose steam locomotives were already at work on colliery tramroads in the North East, and had no doubt about what form of motive power should be used; he said: 'Moving vehicles on rail roads by steam engines can be done cheaper, more certain and twice as expeditiously, as by boats on navigable canals, and neither the repairs of the locks and banks, the want of water, the

summer's heat or the winter's frost, will retard the operation'.

Within a few years he was shown to be, to a large extent, a true prophet: ironically, the Stratford & Moreton Railway was built and worked as a horse line. The first public railway to use steam locomotives was George Stephenson's Stockton & Darlington, over a route long planned as a canal but authorised as a railway a few weeks before the Stratford & Moreton, and opened in 1825.

So the numerous proposals which were made in the late 1820s and the 1830s for improvement of canals, and construction of new ones, were made against a swelling tide of proposals for steam railways, proposals which in steadily increasing number matured into construction.

This competition brought its own

compensations. Just as in later years the steam locomotive was to reach the peak of its development in the 1930s against diesel competition, or the sailing ship in the 1860s against competition by steam, so, in the 1820s and 1830s, canal engineers were stirred to do their finest work.

Canal and rail between London and Lancashire

The corridor within which the battle between canal improvement and railway construction was fought, and eventually for canals heroically lost, was that between London, Birmingham, Liverpool and Manchester. Already then, as now, it incorporated the principal trade routes of England; and the then existing canals, cramped and circuitous and little changed since they had been built forty or fifty years before, were, with the possible exception of the wide Grand Junction and Bridgewater Canals, no longer adequate.

As early as 1821 William James proposed a railway from Manchester to Liverpool, and in 1824 railways were proposed from London to Birmingham, and from Birmingham to Merseyside. The canal companies in response looked at the possibility of improving the route from London to Birmingham and Liverpool.

The first to take action was the Birmingham Canal Navigations: indeed action was probably forced on it as much by the inadequacies of its own main canal as by railway competition. Brindley's original canal of the 1760s had been improved by Smeaton in the 1790s when he lowered its summit, but traffic had since doubled. In Smeaton's day, 100 boats might pass through a lock in twenty four hours: thirty years later it was not uncommon for 200 to do so. In 1824 Telford was asked to survey further improvements.

What he proposed between Birmingham and Tipton was, in effect, a new canal. I was about to add, parallel to the old: but there could scarcely be a less apt description, for whereas the

Birmingham end of the old contour canal was extremely sinuous, the new canal—at the same level—was to be direct, cutting across the old and leaving its loops as branches to left and right. By one long and deep cutting, all locks over the summit were to be avoided by through traffic, and the canal, though intended for narrow boats, was built wide, with dual towpaths and full-width bridges so that boats in each direction could pass one another without hindrance. A new reservoir was to provide water.

These proposals were adopted and the improvements carried out successively over several years, to be followed by a direct line (including a wide, dual-towpath tunnel) between Tipton and Deepfields to bypass the former winding course between those points, and still further improvements after that.

Linked with the Birmingham Canal improvements was a proposal for an improved line from the BCN to Liverpool. The existing route was roundabout and inadequate, by the Staffs & Worcs, Trent & Mersey and Bridgewater Canals to Runcorn and thence down the Mersey. Telford was commissioned by the BCN Company to survey a more direct route: what he proposed was the Birmingham & Liverpool Junction Canal to run from Autherley (on the Staffs & Worcs, close to its junction with the BCN at Aldersley) to Nantwich, where it would join the Ellesmere & Chester Canal and so give access to Ellesmere Port; Liverpool lay beyond, across the Mersey estuary.

Another project which was coming to fruition at this time was the Macclesfield Canal. A canal to serve that town had been considered at intervals for many years; the proposal which was eventually successful originated in Macclesfield late in 1824. The promoters considered and dismissed the idea of a railway and invited Telford to make a survey for a canal. It was to run from the Trent & Mersey north of Harecastle Tunnel (which the T & M had recently obtained powers to duplicate) by Congleton and Macclesfield to the Peak Forest Canal above Marple Locks. This would produce a shorter, though more heavily locked, route from the Midlands to Manchester than the existing route by the Trent & Mersey and the Bridgewater Canals.

There were at this period no less

than three waterway routes between Manchester and Liverpool: the Bridgewater Canal via Runcorn, the Mersey & Irwell Navigation, and the Bridgewater and Leeds & Liverpool Canals via the recently completed L & L branch from Leigh to Wigan. But all these were very busy, and people began to consider the possibility of improving communications still more, by construction of a canal large enough to take small ships inland to Manchester, a proposal that was to crop up in various forms at intervals for many years to come.

A more immediate threat to the existing waterways was the proposal for a railway from Liverpool to Manchester. This got as far as a Bill before Parliament in 1825: but the case for a railway was badly made out and opposition mounted by landowning and waterway interests was successful. The railway promoters, however, decided to try again. A key figure was the Marquess of Stafford, heir of the Duke of Bridgewater. He considered carefully the relative merits of canals and steam railways, asked Telford's opinion, and invested in the Birmingham & Liverpool Junction Canal. But between Manchester and Liverpool he came to the conclusion that a railway was needed, and having done so provided one fifth of the capital. He also provided funds to duplicate the Bridgewater Canal's locks at Runcorn, at that time extremely busy.

In consequence the Act for the Liverpool & Manchester Railway was passed in 1826; George Stephenson became its engineer. The same year the Acts of Parliament were passed for the Birmingham & Liverpool Junction Canal and the Macclesfield Canal. In 1827 a further Act authorised a branch from the B & LJ at Norbury Junction to Newport and onward to join the Shrewsbury Canal at Wappenshall, and in that year also the Ellesmere & Chester Canal obtained powers to build its long-proposed line from Barbridge, near Nantwich, to join the Trent & Mersey at Middlewich. This would give the B & LJ an outlet towards Manchester, although, as its price for the connection, the T & M obtained authorisation to charge stiff compensation tolls at the junction.

The new tunnel at Harecastle came into use in 1827, and for many years both tunnels were used, one for eastbound boats, the other for westbound.

Passage of the Act for the Birmingham & Liverpool Junction Canal dispelled the threat of a railway over the same route. Telford is said to have laughed heartily. Not that he was dogmatically opposed to railways, but he considered them, after investigation, inferior to canals, and felt that the future of steam transport lay in road vehicles. Both opinions were, at the time, justifiable. The practicability of the steam locomotive was not yet wholly proven: the Liverpool & Manchester Railway Company was still considering whether it might be best to work its trains by stationary engines and cables between the rails. Besides, a colossal amount of energy and capital had been put into construction of canal and road networks over the previous fifty years: was yet another means of transport, for which a third network of specialised tracks would be needed, really justifiable?

Telford was the engineer of the Birmingham & Liverpool Junction, consulting engineer of the Ellesmere & Chester. He had surveyed the Macclesfield, but the engineer appointed to construct it was William Crosley, who had previously worked on the Lancaster Canal. These canals, built late in the canal era against a threat of competition, were magnificently engineered, forsaking the contours and taking direct lines across country by deep cuttings and high embankments. On the Macclesfield Canal in particular the quality of the masonry, of bridges and aqueducts, is superb, and to cruise along this canal today, following the lower slopes of a range of high hills, remains an exhilarating experience.

In one respect these new canals were, however, a reversion to old ideas. There had been a time, as I have mentioned, while the Grand Junction was being built as a wide canal, when it had looked as though many other canals in the Midlands might be built wide or widened. But they had not, or not enough of them to make a network of wide canals: by the 1820s the wide faction had lost, and the new canals authorised then, however advanced their engineering in other respects, were built only for narrow boats.

A firm proposal to improve the canal route from Birmingham in the direction of London was made in 1827. The London & Birmingham Junction Canal was proposed, with Telford as engineer, to run from the Stratford Canal (at the end of the long level pound which led from Birmingham) to Braunston, with a branch to Coventry. It was to pass to the north of the valley of the Avon, and so have fewer locks than the Warwick & Birmingham/ Warwick & Napton line, which descended into the Avon valley at Warwick and then climbed out of it again, or the alternative and less heavily locked but longer route via Fazeley.

The immediate effect of this was felt elsewhere. The Oxford Canal Company, alarmed at the prospect of losing a great deal of revenue if the proposed canal were built, decided to do what had often been considered: to improve the northern end of its line. It had this surveyed, and obtained an Act in 1829 to shorten it, cutting off the most exaggerated contour-following loops by construction of a new direct canal over the same general route. The length of the line, between Hawkesbury where it joined the Coventry, and Wolfhampcote between Braunston and Napton, was to be reduced from thirty seven miles to twenty three and a half. As a result of this the London & Birmingham Junction promoters altered their proposals so as to join the Oxford Canal near its northern end, but nevertheless failed to obtain an Act of Parliament.

In 1829, with its line approaching completion, the Liverpool & Manchester Railway Company organised a series of locomotive trials at Rainhill, to establish whether or in what form steam locomotives would be suitable for its traffic. Stephenson's latest locomotive *Rocket* was outstandingly successful, and from then onwards the practicability of steam railways was certain. The Liverpool & Manchester Railway was opened in September 1830.

However it did not, for some years, succeed in carrying more than one third of the total freight traffic between the towns of its name. Its immediate success, to the surprise of its promoters, was with passenger traffic. But the waterways maintained their freight traffic only by cutting rates. This was to become a common story. A curiosity, in modern terms, is that goods traffic on the railway, as on a canal, was conveyed by carriers, at least one of whom ran his own goods trains and also carried by canal.

The following year, 1831, the Macclesfield Canal was opened; and during the early 1830s the Trent & Mersey duplicated the many locks on the descent from Harecastle into Cheshire, to meet the demands of traffic.

Railway surveyors and promoters had been hard at work and in 1833

5/4 Brindley's original Harecastle Tunnel, right, was later duplicated by Telford's new tunnel, left. These are the western portals.

Parliament passed Acts for the first two trunk railways: the London & Birmingham Railway and the Grand Junction Railway. The latter was to run from Birmingham to a junction with the Liverpool & Manchester midway along its length. Construction of both lines commenced.

On the canal front, the Ellesmere & Chester's Middlewich branch was opened in 1833, and in 1834 the Oxford Canal's new line along its northern section was completed. Telford died the same year. A few months later his last great work, the Birmingham & Liverpool Junction Canal, was completed in 1835 after immense difficulties of construction. Improvements at Ellesmere Port, new locks and a new dock and warehouses, largely to Telford's design, were commenced the following year.

Passage of the Act for the London & Birmingham Railway did not immediately discourage the promoters of an improved canal line between these points. As late as 1826 there was a proposal for an improved canal direct from the Stratford Canal to the Regent's Canal, intended to halve the time taken by boats between Birmingham and London. This did prompt the Grand Junction Canal to duplicate the flight of wide locks at Stoke Bruerne to ease congestion there, but otherwise support for a new canal started to diminish as construction of the London & Birmingham Railway progressed.

Of the two trunk railways, the Grand Junction Railway was opened first, in 1837; the London & Birmingham, late in 1838. There was now railway connection between London, Birmingham, Liverpool and Manchester. The railway era had arrived.

Canals and railways elsewhere

While so much had been happening in the London-to-Lancashire corridor, there had been many independent developments elsewhere. The Thames & Medway Canal was opened in 1824; of its length of seven miles, two and a quarter were underground, where it passed through the great Strood Tunnel which had been found necessary to penetrate the ridge between the two rivers of its name. This was the second longest canal tunnel, and it had the largest bore, to take estuary craft. The same year the Earl of Grosvenor installed lock gates at the mouth of a tidal creek off the Thames which he owned. It then became known as the Grosvenor Canal, running northwards for three quarters of a mile from the Thames near Chelsea.

An ambitious project was the Ulster Canal, authorised in 1825, to be over forty five miles long and to link the River Blackwater, which flows into the south-west corner of Lough Neagh, the head of Upper Lough Erne, whence a complex of natural channels and lakes give access to Enniskillen and Belleek.

The Inhabitants of Bolton, Lancashire, had long expected to see their town joined to the canals to the west by construction of a canal to link the Manchester, Bolton & Bury to the Lancaster. After the Wigan-to-Leigh branch of the Leeds & Liverpool was completed this, for reasons mentioned earlier, became unlikely, and in 1825 an Act of Parliament was passed for the Bolton & Leigh Railway, which was to run from the MB & B Canal at

5/5 *Intensive traffic caused the Trent & Mersey Canal to duplicate the locks on the western ascent to Harecastle during the 1830s. This meant also duplicating bridges, at the tails of locks: here, the older lock and bridge are straight ahead, aligned on the centre of the canal, and the newer ones are to the left.*

the Stratford & Moreton Railway (in practice, a horse tramroad) was opened in 1826.

The Aire & Calder Navigation's canal to Goole was opened in 1826 also. During the period of construction the point at which it was to diverge from the existing navigation had been moved slightly, from Knottingley to Ferrybridge, and the company had decided to build docks and a new town at Goole. Radical improvements above Ferrybridge were already being considered, mainly in the form of lateral canals, and an Act was obtained two years later.

In 1827 came a remarkable event: completion of the Gloucester & Berkeley Ship Canal, which had been authorised during the Canal Mania thirty years before. The line had been shortened slightly to enter the Severn at Sharpness, but the canal was still over sixteen miles long and large enough to take vessels of 600 tons capacity to Gloucester which became, and remains, Britain's furthest inland port for seagoing ships. Further to the south-west, the same year, the Bridgewater & Taunton Canal was opened.

The Exeter Canal had been dredged and straightened in the early 1820s and it was then decided to extend it so as to enter the estuary lower down, at Turf, so that it might take small ships, and to build a new basin at Exeter. The engineer was James Green, who had been engineer of the Bude Canal and was to be associated with the abortive project of 1836 for a new canal from Birmingham to London. The extended canal was opened in 1827 and the basin in 1830.

In 1828 the Bolton & Leigh Railway was opened, with a steam locomotive in use and two inclined planes. By then the Liverpool & Manchester Railway was being built, and would pass a few miles south of Leigh. In 1829 an Act was obtained for the Kenyon & Leigh Junction Railway, to extend the line of the Bolton & Leigh to a junction with the Liverpool & Manchester. This was built and so the Bolton & Leigh Railway, originally proposed as a link between canals, became in effect a branch of the Liverpool & Manchester Railway. It never did connect with the Manchester, Bolton & Bury Canal, although that company's engineer was among its shareholders.

In 1829 a tramroad or railway was opened to connect a colliery at Shut

Bolton to the Leeds & Liverpool at Leigh. Another railway or tramroad authorised in 1825 to link canals was the Cromford & High Peak Railway. It was the outcome of many years of proposals to link the Peak Forest Canal with canals east of the Pennines and, by connecting with the Cromford Canal, was to provide a route from Manchester to the Nottingham area,

and perhaps even London also. The Macclesfield Canal at this time was still at the proposal stage. Steam power, both mobile and stationary, was authorised for the Cromford & High Peak Railway, and with a length of thirty three miles and six inclined planes it was the most ambitious rail link built to form part of what was primarily a canal route. Further south,

5/6 Shelmore Embankment, on Telford's Birmingham & Liverpool Junction Canal, towers above a canalside farmhouse. Grid reference SJ 795225.

End, west of Dudley, with the Staffs & Worcs Canal. Of purely local importance at the time, it is of interest here because it used from the first the locomotive *Agenoria*, which is now an exhibit in the National Railway Museum, York: the only locomotive from a canal tramroad to survive.

The Thames & Medway Canal had become busy enough for a passing place in Strood Tunnel to become essential. This was made in 1830 by opening out a short length of tunnel, where it was closest to the surface. There was no hint yet of what was to become of the canal through the tunnel, of a course which the Manchester, Bolton & Bury Canal was

already considering: conversion of the canal into a railway.

The MB & B obtained an Act in 1831 which changed the company's name to the Company of Proprietors of the Manchester, Bolton & Bury Canal Navigation and Railway, and authorised it to make a railway 'upon or near' the canal, closing the canal where it was needed for the railway. This met sufficient opposition from canal users to oblige the company to seek a second Act a year later authorising them to build the railway on a separate course and keep the canal open.

Elsewhere in 1831 the Cromford & High Peak Railway was opened. It was

followed in 1834 by the Leeds & Selby Railway, which terminated at Selby beside the tidal River Ouse. The following year the improved and shortened line of the Aire & Calder Navigation was completed to Leeds. In 1835 came the Act of Parliament for another great trunk railway line, the Great Western Railway, from London to Bristol.

The Manchester, Bolton & Bury company having been thwarted of its plan to build a railway on the site of its canal, the fate of being the first canal to be converted into a railway fell to the Croydon Canal. The London & Greenwich Railway was being built in the mid-1830s and, before it was even complete, the London & Croydon Railway was authorised in 1835 to run from a junction with it to Croydon. It was intended that this line should be built on the bed of the Croydon Canal and, since opposition was less strong in Surrey than in Lancashire, the canal was purchased by the railway company and closed in 1837, after a brief life of twenty eight years. The railway was opened two years later: the present-day West Croydon station occupies the site of the canal basin.

Since the railway was in fact laid out on a more direct line than the canal, some sections of the latter were not used. Traces of one of these can still be seen in Betts Park, Anerley (grid reference TQ 347697), where a length of the canal has become a rather elongated ornamental lake, equipped with fountains, lined with concrete and, as when I saw it recently, drained in winter. A row of miniature trees set back from the canal appears to mark the site of the hedge which must once have lined the towpath; to the north the canal enters a shallow cutting to be brought to an abrupt stop by an embankment across it which carries a road—no trace of a bridge remains evident—and to the south the dry course of the canal is clear for a few yards where it provides an entrance to the park from another road. Around are large Victorian villas interspersed

5/7 This building at Whaley Bridge was the interchange point between the Peak Forest Canal and the Cromford & High Peak Railway. Railway tracks entered the two side bays through the doors shown; the canal, approaching the far end of the building, occupies the centre bay and emerges for a few feet at this end. Under threat of demolition when this photograph was taken in 1973, the building has since been reprieved. Grid reference SK 012816.

5/8 (right) The course of the Croydon Canal in Betts Park, Anerley (grid reference TQ 347697). In the background, it has become an ornamental lake, drained at the time the photograph was taken; in the foreground, a pathway curves along its bed.

with small modern houses; Southern Electric trains rumble past nearby.

The Manchester, Bolton & Bury company's railway was opened in 1837, and the same year the Lancaster Canal Company sold its tramroad at Preston to the Bolton & Preston Railway Company, which intended to use the course of the tramroad for its own line, but in the end did not, and so had to maintain the tramroad for the benefit of the canal. Many other canal tramroads were later upgraded into railways, particularly in South Wales.

While this was going on, canals elsewhere were still being extended. The Grand Western finally reached Taunton in 1838: not by barge canal as originally planned, but by much smaller tub-boat canal with an inclined plane and several vertical lifts which raised and lowered boats between the pounds. The proposed line to Exeter never was built.

The Aire & Calder completed its improved line to Wakefield in 1839, reducing the distance from Castleford to Wakefield from $12\frac{1}{2}$ miles to $7\frac{1}{2}$. This navigation now included more artificial canal than natural river.

At this period the Peak Forest Canal, alarmed at the effects on its trade of the opening of the Grand Junction Railway and the London & Birmingham Railway, approached all the other canal companies on the route from Manchester to London in an attempt to get concerted action to reduce tolls and meet competition. It was only partially successful, for canal companies from long experience were too suspicious of one another to combine now against a common threat. With the railways exerting themselves to gain traffic, canal traffic, particularly merchandise which benefited from railway speed, started to fall away, and canals and canal carriers found they could only maintain traffic by cutting tolls and rates.

CHAPTER 6

AFTER THE CANAL ERA

The outline

In the very early days of steam railways, canals were still competitive for carriage of freight. After the St Helens & Runcorn Gap Railway, for instance, had been opened in 1833 from St Helens to Widnes in direct competition with the St Helens Canal (or Sankey Brook Navigation), it was the railway, not the canal, which found itself struggling to survive. The canal had, while the railway was being built, been extended down to Widnes to avoid the difficult navigation of the upper estuary of the Mersey. When the two concerns eventually merged in 1845, the canal, not the railway, was still very much the more profitable undertaking.

So railway companies, or their promoters, had good reason to attempt to obtain control of competing canals. Then, as railways grew and prospered, canal shareholders began to see their profits dwindle, and so to be more and more willing to sell out to railway companies, or to lease their canals to them.

One of the side effects of railway competition was that, to economise, boatmen started to take their families to live on board narrow boats. This avoided the cost of a house on shore and provided a crew as well.

The outline of canal history after the canal era can be told briefly. There was a period during the 1840s and 1850s when a great many canals came under railway control. In England, scattered across the canal network, they were sufficient to split apart those canals which remained independent. Not many canals were actually converted into railways, but railway companies did tend to neglect their canals and to discourage traffic, despite legislation to the contrary. There were exceptions, particularly where a railway-controlled canal served canalside industries not served by rail. But old industries died away and new ones,

coming along, often sought sites which were rail- rather than canal-connected.

In the later part of the nineteenth century some canals which had remained independent, in agricultural areas, closed down, no longer able to compete. But there came at the same period a reaction against railways and in favour of waterways. Some canals were enlarged, some new canals were built. People were aware that canal development had continued on the Continent, despite construction of railways.

This mild revival of canals was overtaken in the twentieth century by the marked revival of road transport which followed the introduction of motor vehicles. Road transport over long distances had earlier been hit by rail competition far harder than canals had been, but its revival led eventually to the virtual collapse of freight carrying on small canals in the British Isles during the 1960s. By small canals I mean wide canals as well as narrow: only on much larger waterways does freight traffic survive in any quantity. Many small canals, however, continue to be maintained in navigable order, partly for pleasure cruising, partly as amenities, and partly because it would be extremely expensive to eliminate them.

Passenger traffic

On looking at the history of canals since the canal era in more detail, however, it becomes clear that there were many exceptions to the general pattern. Canal passenger traffic is an instance. Almost invariably, once a competing railway had been opened, canal passenger boats were withdrawn. The Grand Canal in Ireland carried over 100,000 passengers in 1837; in 1852 it carried none. Railways

competing with the Paisley Canal were opened in 1840 and 1841; canal passenger traffic ended in 1843. But some of the boats were taken south and used to operate a passenger service over the Peak Forest and Ashton Canals which connected with railway trains at Guide Bridge, and later at Dukinfield. This service continued into the 1850s at least. Horse-drawn passenger boats seem to have lasted as long as anywhere on the Bridgewater Canal, where they were still running in the 1860s.

On the Forth & Clyde, horse-drawn passenger boats were replaced in 1852 by a passenger-and-cargo steamer; it ran until 1883. Ten years later, steamers were reintroduced for pleasure traffic and continued until the outbreak of the Second World War. That same occurrence caused the end of passenger steamers on the Caledonian Canal, which had had a steamer service ever since it opened: in later years tourist traffic was more and more in evidence. The passenger service on the Crinan Canal was taken off in 1929 and replaced by a motor bus. It had for many years been the preserve of the little canal steamer *Linnet*, since seagoing passenger steamers had become too big to pass through the canal: *Linnet* connected with them at each end. A year-round service of passenger steamers had been introduced on the Gloucester & Berkeley Canal, which had no immediate rail competition, in the 1850s, and lasted until 1932, after which it too was replaced by motor buses.

Railway Mania

Once the first trunk railways were open and seen to be successful, there developed in Britain a public enthusiasm for promotion and construction of railways which surpassed anything comparable in the Canal Mania, and

was known as the Railway Mania. During the peak year of 1846, no less than 272 Acts of Parliament for new railways were passed. It was during this period that many canals passed under the control of railways, sometimes while the latter were being promoted and sometimes after they were open.

The history of railways, to a greater extent than that of canals, was one of periodical amalgamations of small concerns into larger ones, so that canals which had at first become associated with local railway promotions tended to become eventually the unwanted children of ever-greater amalgamated companies. The Manchester, Bolton & Bury for instance, already a railway-and-canal company, amalgamated with the Manchester & Leeds Railway in 1844, so that the canal passed successively through the ownership of those two companies followed by the Lancashire & Yorkshire Railway and in turn the London, Midland & Scottish. In 1841 the Royal Canal was sold to a newly-formed railway company, and the Huddersfield Canal agreed to sell out to the promoters of a competitive railway line which was proposed.

Such sales required the authority of Acts of Parliament and when the Ashby Canal proposed to sell out to the Midland Railway in 1845, opposition by the Oxford and Coventry Canal Companies resulted in clauses being included in the Act to ensure not only that the canal should be kept in good order but also that tolls should not be raised. The canal was therefore able to survive railway ownership comparatively unscathed for many years. Likewise the Trent & Mersey was purchased by the North Staffordshire Railway Company in 1845 and, with the exception of the Froghall–Uttoxeter section which was converted into a railway, it was run by the railway company with some vigour because it served districts beyond the reach of its railway system. The same applied to the Ashton, Peak Forest and Macclesfield Canals which severally fell to the Manchester, Sheffield & Lincolnshire Railway.

Least fortunate were those canals which passed under the control of that otherwise admirable company, the Great Western Railway. One of them was the Kennet & Avon which sold out to the GWR in 1851 in return for a guaranteed dividend in perpetuity.

6/1 Paddle steamer Gondolier *carried passengers and mail along the Caledonian Canal. She was built in 1866 and depicted like this when new; she lasted until 1939. In 1970 canal staff were still referring to her habitual mooring at Banavie as the Gondolier Quay, and may well still do so.*

The policy of the GWR was that traffic should go by rail: by way of encouragement, dredging and maintenance of the canal were neglected, tolls were raised and night traffic prohibited.

The Stratford Canal, too, eventually fell to the GWR, having been purchased by a constituent company in 1859 after a complex series of negotiations which commenced in 1845. Other canals which passed into railway ownership between 1847 and 1854 included the Ripon, Pocklington, Cromford, Nottingham and Edinburgh & Glasgow Union Canals. The Fossdyke was taken over by the Great Northern Railway on a long lease, and the Grand Western Canal was leased in 1853 to a railway company which later exercised an option to purchase. Its eastern section, still fairly new, from Taunton to Lowdwells, was closed in 1867. The Forth & Clyde Canal, after absorbing the Monkland Canal in 1846, dallied with various railway companies for years and was eventually bought in 1867 by the Caledonian Railway Company.

The Bridgewater Canal's trustees purchased their old rival, the Mersey & Irwell Navigation, in 1846, and maintained their independence until 1871. Then, because railway control of canals was becoming unpopular, the canal and river navigations were purchased jointly by the chairmen of the Midland and MS & L Railways, who then formed the Bridgewater Navigation Company to own and run them.

A different route into ownership by a railway was taken by the Monmouthshire Canal. In the 1840s it obtained powers to build railways, upgrade its tramroads into railways, and call itself the Monmouthshire Railway & Canal Co. In 1865 it absorbed the Brecon & Abergavenny Canal, and the whole undertaking was amalgamated with the GWR in 1880. The River Don absorbed the Dearne & Dove and Stainforth & Keadby Canals and a local railway in the 1840s: here again all were later taken over by a railway company, in this case the Manchester, Sheffield & Lincolnshire.

Of those canal companies which did convert all or part of their lines into railways, the most remarkable was the Thames & Medway. In 1844 it built a single track railway alongside its canal from Gravesend to Strood. Through Strood Tunnel this meant that the track was carried partly on what had been the towpath, partly on a continuous bridge or staging built out over the water. So the canal was restricted in width, but not blocked entirely. Passenger trains ran on this line, connecting at each end with steam boats on the Thames and the Medway, for there were as yet no other railways in the vicinity. It was opened in 1845, but the same year the company decided to sell out to the South Eastern Railway Company. The canal through the tunnel was subsequently closed and a double track railway laid: this now forms part of British Rail's North Kent line. The rest of the canal continued to be used by a little local traffic for many years. Much of it can still be seen, reedy and overgrown but still in water, and the basin at Gravesend is used for moorings for pleasure craft. At Strood the basin remains in existence but out of use.

In London, the upper part of the Grosvenor Canal was sold in 1858 and filled in to become the site of Victoria Station. However a short section of the canal, including the impressive entrance lock from the River Thames, is still in use: the traffic is lighters laden with rubbish. In Scotland, the Paisley Canal was bought by a railway company in 1869 and much of its course was subsequently used for a railway, although there also a short section still exists, out of use.

The BCN and the Shropshire Union

Two important canals which came partially but not wholly under railway control were the Birmingham Canal Navigations and the Shropshire Union Canal (of which more shortly). In 1846 the BCN absorbed the Dudley Canal and the combined system was leased to the newly-formed London & North Western Railway (its principal constituents were the London & Birmingham and the Grand Junction Railways). Terms of the lease were that the railway company would guarantee a dividend of four per cent on the canal company shares: so long as the canal company paid as much or more from its own resources, it was to remain in control with only a few restrictions in favour of the railway company. The guarantee was not called upon until late in the 1860s.

The area served by the BCN contained, in enormous quantity, factories and industries which had been placed beside canals so as to use them for transport. To serve them all by railway sidings was impracticable. The BCN was therefore developed further for local transport and many railway interchange basins were built. Goods were transferred there between rail and canal; collections and deliveries, by boat, were made between them and canalside industries. Boatage, cf cartage, was the term used. Interchange basins linked the BCN not only with its overlord the LNWR, but also with the Great Western and Midland Railways.

The BCN improvements which had commenced with the new main line had been continued into the early railway era: the most important at this stage was the Tame Valley Canal, a new eight-and-a-half-mile modern-style canal opened in 1844 which enabled boats travelling between the Birmingham & Fazeley Canal and the Wednesbury area to avoid the B & F's congested flights of locks up into Birmingham itself. The improvements continued further into the period of the railway lease (but, still, canal control). The most important was construction of Netherton Tunnel, one and three quarters miles long, which duplicated the busy and restricted Dudley Tunnel and lies a few miles to the east. It was the last canal tunnel to be built, and it was made wide enough for narrow boats to pass one another, with dual towpaths and gas lighting. It was opened in 1858.

In 1845 the Ellesmere & Chester Canal absorbed the ten-year-old Birmingham & Liverpool Junction. The Railway Mania was approaching its peak and the combined company then obtained powers to purchase the Montgomeryshire Canals, convert many of its canals into railways, build other railways, and change its name to the Shropshire Union Railways & Canal Company. Construction of railways started with the Stafford-to-Shrewsbury line (over new ground: not a canal conversion). Then, in 1846, the LNWR offered to take a perpetual lease of the Shropshire Union system, in return for a guaranteed dividend, with management by a joint committee drawn equally from the two companies. This was accepted, and the SUR & C Co. became another satellite of the LNWR. The Stafford-Shrewsbury railway was completed but it built no further railways; the canal system was worked quite energetically as it gave access to places the LNWR did not otherwise reach. Parts of the Shropshire Canal, which tub-boat canal had been leased by the SUR & C Co., and which suffered from subsidence, were closed in 1858 and used for the LNWR branch railway to Coalport.

Some waterways which were leased by railway companies later reverted to their owners. These included the Calder & Hebble Navigation and the Rochdale, Leeds & Liverpool and Grand Canals. The Lancaster Canal turned the tables when, for several years during the 1840s, it leased the Lancaster & Preston Junction Railway. In 1864, it leased its northern section to the LNWR and its southern section to the Leeds & Liverpool Canal. The LNWR later purchased the canal outright, and the L & L then continued to rent its southern end from the railway company. The connecting tramroad had been closed, isolating the northern section from the rest of the canal system, and the southern section, built for the north-south line of the Lancaster, came to be thought of as part of the Leeds & Liverpool's east-west line.

Rail competition hit canal carriers as hard as canal companies. Pickfords asked for toll reductions in 1844 and, failing to get them, ceased to carry by canal in 1846. To counteract this tendency Parliament found time, amid all the railway legislation, to pass the Canal Carriers Act 1845 which authorised all canal companies to carry goods with their own boats. Companies which then established carrying fleets included the Grand Junction, the Leeds & Liverpool and the Grand Canal. The Birmingham & Liverpool Junction had already obtained powers in 1842 to set up a carrying department and this was taken over and expanded by the Shropshire Union.

River improvements

On the River Thames, many pound locks were built from 1809 onwards. The last of the series, Bray Lock, was completed in 1845. The 1840s saw the start of great improvements, by construction of locks and dredging, to two great rivers: the Severn and the Shannon. The Caledonian Canal was closed between 1843 and 1847 for extensive repairs amounting almost to rebuilding.

In Ireland, the Ulster Canal was completed in 1841, and in 1846 the Board of Works, a Government body, started to build the Ballinamore & Ballyconnel Canal, intended to link Upper Lough Erne with the Shannon Navigation at Leitrim village. By doing so it would, with the Ulster Canal and the Lagan Navigation, complete a waterway from Belfast to the Shannon. The Shannon improvements, mentioned above, were contemporary.

The B & B Canal was intended in part for land drainage, for which it may have had some use, but as a navigation it was a flop: for when it was opened in 1860, no traffic arose. On one occasion only was it used commercially from end to end, when a tug and coal barges passed through. Pleasure craft used it occasionally, but after 1869 it became disused and, eventually, derelict. Traces of the navigation works survive—substantial stone bridges and gate-less lock chambers through which water flows unimpeded—and can still be seen, particularly at Leitrim and near Ballinamore (Co. Leitrim), among other places. There have for some years been proposals to rebuild and reopen the canal for pleasure cruising, which has become popular on both the waterways with which it connects.

Farther to the west the Board of Works started to construct canals in the early 1850s, to connect Lough Corrib with the sea at Galway, and with Lough Mask. The canal at Galway, named the Eglinton Canal after the lord lieutenant who opened it in 1852, remained navigable until 1955 when its life-expired swing bridges were replaced by low-level fixed bridges. When I looked at this canal in 1969 its two locks appeared to be complete though inaccessible, for the bridges had insufficient headroom for a rowing boat: and in this condition I understand it still remains.

The attempt to build the canal between Lough Corrib and Lough Mask became notorious. When it was completed, with four locks, the canal failed to hold water owing to the porous nature of the limestone through which it was built, and the works were eventually abandoned. I have heard it suggested that a cure could have been found, but that funds, intended primarily for famine relief, were exhausted. Monumental remains of lock chambers and cuttings through rock can be seen at Cong, Co. Mayo.

One of the first canals to be closed was the ancient Stamford Canal. It had never been linked to the main canal system and by the 1820s was in poor condition. From 1846 Stamford was served by railway and in 1863 traffic on the canal ceased. The Corporation of Stamford then attempted to put the canal, together with the two locks further down river, up for sale by auction in twenty four lots. This sale was contested on the grounds that the Corporation was not the owner of the canal but merely trustee for the public. Particularly concerned were the inhabitants of Market Deeping and Deeping St James, villages lining the banks of the Welland immediately below the end of the canal. Their concern was not directly about navigation, however, but that if the two river locks were sold the water levels would not be maintained, reducing the river to a stagnant stream in summer and rendering useless conduits leading from it for water supply. Stamford Corporation does appear to have disposed of most of the canal subsequently, and a committee was formed at the Deepings to maintain their two locks. It seems to have been effective, for during the course of a recent visit I noted that the depth of the river is still maintained by weirs at the sites of the two locks, and at one of them, Lower Lock, Deeping Gate (grid reference TF 148096) the lock chamber (or *pen*, in local parlance) remains intact, although the gates have long since gone. Elsewhere there are still substantial traces of the canal, both in water as at grid reference TF 132098 where it shows up as a weed-covered channel among the trees beside the main road, and drained, for instance at grid reference TF 066069 where the dry bed enters a noticeable cutting, although a road bridge has been filled in. The villages called the Deepings lie at the edge of the Fens, but nevertheless have many old stone houses suggestive of water transport from quarries to the west.

The Wey & Arun Junction Canal was never very profitable, and once a competing railway (the Horsham–Guildford line) was eventually opened, in 1865, it receipts fell by half. The inevitable consequence was closure of the canal in 1868. The northern part of the Arun Navigation lasted only another twenty years.

Today, 110 years after closure, there are still some remarkably substantial traces of the Wey & Arun, as I observed during a recent visit which was all too brief. To begin with, the northernmost 260 yards or so are still navigable: this length appears as a quiet backwater of the River Wey (grid reference SU 997465) but its straightness confirms its artificiality, as it runs between trees and orchards, lined with moored pleasure craft, as far as a culvert in a brick wall supporting a main road. The arch of the former bridge, now bricked up, is clearly visible.

The parts of Surrey and West Sussex through which the canal passes are a bosky region: they were particularly so at the end of May when I was there, and it would have been easy to miss the canal. Where the A281 Guildford-to-Horsham road crosses it near Alfold Crossways (grid reference TQ 041367) it seems, from the map, that the canal must be clear enough. Yet one can drive past this point without noticing it, for the road across the canal is level, and the canal bed hidden behind hedges. A by-road diverges to the east, and only when I had turned into it did I observe that an old canal bridge survived, screened from the main road by a hedge. No doubt it once carried the main road, which had been diverted and levelled out: for it revealed itself as

a brick humped bridge of the traditional canal type, though much overgrown. The canal beneath retained a little water in its bed, but was overgrown with trees. So was the canal elsewhere: in the period of over a century since the canal closed, innumerable trees, in this thickly-wooded country, have taken root beside the canal or on its bed, grown up, matured and eventually died and collapsed into it.

The course of the canal was equally invisible where crossed by the B2133 road at Loxwood (grid reference TQ 041311), yet to step through a small gap in the thick hedge on the road's east side was to have its course revealed: a damp depression in the ground, bright with wild flowers, stretching invitingly ahead. The one-time towpath has become a footpath and, by following it for about 400 yards, I came to the remains of Brewhurst Lock half-hidden beneath the jungle of West Sussex. The masonry of its chamber appeared to be almost complete, but from the sill,* where once the top gates were, a full-grown ash tree sprang; another, its trunk splitting into two, was rooted in the masonry of the chamber, with many smaller trees and bushes and creeping plants which festooned the lock sides. Remains of the lower gates still lay in the bed of the canal but were inaccessible because of overgrowth; and what at first glance appeared to be two steel piles, rising on either side of the gap which represented the top gates, were revealed on closer inspection to be iron quoins† into which the edges of the top gates had fitted. The masonry behind them had fallen away.

* The terms used for components of a lock are explained on page 126.

†The quoin, in the context of a lock, is the curved recess into which the heelpost of a lock gate (that is, the edge at which it pivots) is pressed by pressure of water.

6/2 *The remains of Brewhurst Lock, Wey & Arun Junction Canal, in 1978, 110 years after closure. The tree is rooted in the sill of the top gates, a cast iron quoin stands on the right, creepers on the left trail down the chamber side, and remains of the lower gates can still be seen in the damp bed of the canal. Grid reference TQ 043311.*

To see a canal like this, over a route which would be useful to pleasure craft, is to wonder about the possibility of its being restored for navigation. The Wey & Arun Canal Trust has restoration as its purpose: it was founded in 1973, successor to a society formed earlier, all part of a nationwide movement for restoration of disused waterways which gets fuller mention later. So far as the Wey & Arun is concerned, the task of restoration is vast, but the appeal of it is understandable. The trust is tackling it in short lengths and sections, rather than preparing a detailed overall programme, with costs, lest this put people off. It is a wise course.

A most useful publication, for those wishing to visit the canal is the trust's booklet *Wey-South Path*. This gives details, with maps, of a walking route from Guildford to Amberley, alongside the courses of canal and Arun Navigation where these are accessible to the public, and as close as possible to them where they are not.

An example of the results of the trust's activities could be seen at Rowner Lock (grid reference TQ 069271). Here the trust, starting with a lock chamber only a little less derelict than the one just described, had restored and repaired its walls and sills, rebuilt the parapets of a bridge at the tail of the lock, and installed new top gates. New bottom gates were being made, and water had been restored to navigation level in the section from Rowner south to Newbridge.

Several other canals were closed or became disused as a result of rail competition: brief mention of a few must suffice. In South Wales, railway competition came late. The first railway, rather than tramroad, in the South Wales valleys was the Taff Vale, opened in 1841, and the canals flourished until the 1860s. Then as old industries closed down and new ones sought rail connection, canal traffic collapsed and the canals were little used after the turn of the century. The Ripon Canal was disused after 1892, and the Thames & Severn was closed between Chalford and the Thames in 1893. It had been in a poor state, physically and financially, for years, and in 1883 the Great Western Railway had acquired a controlling interest as a defensive measure lest Sapperton Tunnel be used for a competing railway.

Closure was of sufficient concern to connected waterways and local authorities for them to form a trust to take over the canal, to which the GWR was agreeable on condition railway conversion did not take place. The canal was reopened in 1899, transferred to Gloucestershire County Council in 1901, and lingered on until after the First World War.

About the turn of the century much of the remaining part of the Shropshire Canal ceased to be used, and the last traffic passed over the summit of the Huddersfield Canal at the same period. The Bude Canal also became disused at this time.

Regulation of railways

The hindrances to traffic which occurred on railway-owned canals such as the Kennet & Avon resulted in an important piece of legislation in 1873. This was section 17 of the Regulation of Railways Act of that year, and is worth quoting in full:

'Every railway company owning or having the management of any canal or part of a canal shall at all times keep and maintain such canal or part, and all the reservoirs, works and conveniences thereto belonging, thoroughly repaired and dredged and in good working condition, and shall preserve the supplies of water to the same so that the whole of such canal or part may be at all times kept open and navigable for the use of all persons desirous to use and navigate the same without any unnecessary hindrance, interruption or delay'.

This was forthright stuff, a strong reinforcement for the existing public right of navigation (and one could do with it in relation to British navigation authorities today. It continues to apply to CIE vis-à-vis the Grand Canal in the Irish Republic, which inherited this law and, unlike Britain, has not repealed it). Its effect, however, was curious, for in the long run it ensured the survival of several railway-owned canals long after they had become intrinsically uneconomic and would have closed had they remained independent. Ownership by a railway, therefore, was not all bad. Probably its worst feature was not so much neglect

as an unwillingness, except in special circumstances, to promote or improve a canal. One of the most regrettable instances is that of the Forth & Clyde Canal. By the turn of the century the owning Caledonian Railway was giving as a cause for declining traffic a general increase in the size of coastal ships so that they could not pass through the locks. It made, however, no attempt to enlarge them, preferring to concentrate on its main business of carriage by rail. Had the locks and the canal been enlarged I am convinced that the Forth & Clyde would have survived as a busy commercial waterway, instead of one which is today disused and half-dead.

The comparison is the Aire & Calder, which did remain independent, and did enlarge, and is now Britain's busiest commercial waterway.

The Lancashire & Yorkshire Railway reached Goole in 1848, and the Aire & Calder Navigation built railway facilities for it within the dock area. Later these were transferred to the railway company. But this was as far as the Aire & Calder went towards rail take-over. In 1852, twenty-two-year-old W. H. Bartholomew succeeded his father as engineer of the navigation, with important consequences. The Aire & Calder had introduced steam tugs in 1831, and in the 1850s new tugs were built. It became the practice for them to haul keels in trains, and in 1859 the company started to lengthen its locks, so that larger vessels and longer trains of keels—up to ten of them—might pass through.

In 1864 came the most important development in canal-boat design since the narrow boat. To meet rail competition, Bartholomew designed and patented a system of compartment boats for coal. Each boat was to carry twenty five tons, and they were to be propelled in articulated trains of about six by a steam tug. At the head of the train was a bow compartment, not cargo-carrying, and a system of wires enabled the whole train to be curved in order to steer it. At Goole, a hoist lifted each boat out of the water and tipped its contents into a waiting ship. Technically the system was later slightly modified, and is referred to again in chapter eight; commercially it was a great success.

From 1863 for twenty five years the Aire & Calder leased the Calder &

6/3 Sir Edward Leader Williams: designer of the Anderton Lift and engineer of the Manchester Ship Canal.

down chutes to larger craft on the river below. In 1872 it was decided to build a boat lift to transfer boats or barges between the two waterways. The structure was built by the Weaver Navigation Trustees. It was originated and was in outline designed by their engineer Edward Leader Williams; during its construction however he was lured away to become engineer to the newly-formed Bridgewater Navigation Company. The lift was completed in 1875; as built it operated hydraulically, and boats were raised and lowered in two tanks of water which counterbalanced.

The Manchester Ship Canal

By the 1880s Manchester was a city in decline, its cotton industry in the depths of a slump. Prime causes were high railway freight rates and, in particular, high port charges at Liverpool, through which Manchester's exports and imports passed. The possibility of a ship canal to Manchester had been considered at intervals, and detailed plans were prepared in the late 1870s. Daniel Adamson led the promoters and, after a long battle in and out of Parliament, the Act for the Manchester Ship Canal was passed in 1885. The engineer was E. Leader Williams, who had earlier left the Bridgewater company; the canal was to be 36 miles long, from Eastham on the west shore of the Mersey estuary to Manchester, with locks 600 feet long and 65 feet wide.

It was to be built along the west and south sides of the estuary and then, largely, along the course of the Mersey & Irwell Navigation; this meant that the ship canal company needed to purchase that navigation and so had to buy the Bridgewater Navigation Co. Ltd, acquiring the Bridgewater Canal as part of the bargain. Enthusiasts for serendipity may care to note that the cheque made out to complete the purchase in 1887 was, at £1,710,000, the largest which had been presented up to that date. Total cost of construction of the ship canal was some £15 million: raising this sum was not easy and eventually a substantial amount was provided by Manchester City Council on condition that it

Hebble Navigation, and lengthened some, but not all, of the locks. But it lost money on the C & H, and when the lease expired, allowed it to lapse. The Aire & Calder itself, however, became from the 1880s onwards extremely prosperous.

Aire & Calder progress was paralleled elsewhere. The Goucester & Berkeley Canal brought steam tugs into service to haul sailing ships and barges. Then, finding that its entrance at Sharpness was overcrowded and too small for steamships, it set about construction of a new large entrance and docks. These, the Sharpness New Docks, were opened in 1874. Simultaneously, with a view to promoting Birmingham traffic, the company pur-chased the Worcester & Birmingham Canal Company, which by then was in receivership, and changed its own name to the Sharpness New Docks & Worcester & Birmingham Navigation Company.

It will be remembered that in the early days of canals it had been decided that the western end of the Trent & Mersey should join not the River Weaver but the Bridgewater Canal. However the Weaver had become a busy navigation and the T & M passed close to it, although at a level some fifty feet higher, at Anderton, Cheshire. An important traffic on both waterways was salt and the practice had grown up of transferring this commodity at Anderton from boats on the canal

The south portal of Tardebigge Tunnel, near Bromsgrove, Worcester & Birmingham Canal. This tunnel was built wide, in anticipation of use by wide vessels from the River Severn, and completed in 1810. But economy then dictated construction of narrow locks down to the river. Lack of a towpath meant that horsedrawn boats had to be legged through; later, tugs were used. One of them, motor tug Worcester, *survives as an exhibit at the Boat Museum, Ellesmere Port.*

'The Tunnel', Fenny Compton, Oxford Canal, north of Banbury. James Brindley and his assistants knew little of making cuttings, and preferred to tunnel, so that in several places their canals passed through tunnels which were very close to the surface. The Oxford Canal's tunnel at Fenny Compton, built in the 1770s and nearly half-a-mile long, was one such: to eliminate the obstacle it presented to traffic, it was opened out into a cutting in the 1860s. The name 'The Tunnel' survives, however, even to the extent of appearing in print on the Ordnance Survey map.

Turnover bridge over the Macclesfield Canal at Marple, Stockport. This graceful design originated from the practical need to enable a boat horse to cross from one side of the canal to the other without having its tow-line detached, which meant that it had to walk beneath the bridge first, before turning back up onto it. Through towns, wharves were commonly on the downhill side of this canal, so the towpath was transferred to the other bank to avoid the obstacles of moored boats and wharf frontages. Narrows in the foreground are the chamber of a stop lock which once separated the water of the Macclesfield Canal from that of the Peak Forest Canal beyond the bridge.

should appoint a majority of the canal company's directors. The canal was completed at the end of 1893 and formally opened by Queen Victoria in May 1894.

The ship canal in due course prospered and continues to do so, so that its subsequent history, though fascinating, can scarcely be considered to be within the scope of archaeology. The effect of its construction on other waterways is relevant, however. It intercepted all the waterways entering the south side of the Mersey—the Shropshire Union, the Weaver and the Bridgewater Canal, and these subsequently connected with the ship canal instead. The uppermost section of the Mersey & Irwell Navigation, above the new docks built at Manchester, continued to be maintained, giving access to the Manchester, Bolton & Bury and Bridgewater Canals. The main problem was at Barton, where the ship canal occupied the former course of the Mersey & Irwell Navigation. This meant that Brindley's aqueduct, with insufficient headroom for ships, had to be demolished. It was at first proposed that the Bridgewater Canal should cross the ship canal by a high-level aqueduct approached by a pair of Anderton-type lifts; then to avoid the delays inherent in such an arrangement, Leader Williams designed the remarkable swing aqueduct, comparable to the ship canal's swing bridges, which was built and continues to be used. There is more about it in chapters seven and ten.

The Sheffield & South Yorkshire formed

Construction and completion of the Manchester Ship Canal gave a boost to growing contemporary enthusiasm for water transport, of which the efforts made at this time to secure the continued existence of the Thames & Severn were one facet. Another example was the incorporation by Act of Parliament in 1889 of the Sheffield & South Yorkshire Navigation Company. Its intention was to acquire from

the Manchester, Sheffield & Lincoln-shire Railway the waterways between the Trent at Keadby and Sheffield, and to enlarge them so as to take 400-ton barges. Such a scheme has a familiar ring to present-day students of canals: if the SSYN promoters had had foreknowledge that, ninety years later, the Doncaster-to-Rotherham part of their proposal would only just have been authorised, and its onward projection to Sheffield forgotten, their enthusiasm might have been a little less strong. A more immediate problem was that the MSLR was not a willing vendor of its waterways (it had opposed the Act); in consequence it was 1895 before the new company succeeded in taking over its waterways, and then it found itself, because of difficulty in raising money, with half of its directors appointed by the railway.

The improvements that it was at first able to make to its line were therefore minimal. But there was one important outcome: construction of the New Junction Canal. This five-and-a-half-mile canal was promoted in 1890 jointly with the Aire & Calder to link the SSYN at Bramwith below Doncaster with the A & C about seven miles from Goole. It was eventually opened in 1905 and enabled vessels from the SSYN to reach Goole docks, and, more important, compartment-boat trains from the A & C to reach the SSYN.

An abortive proposal of the period was for a ship canal to run from the Weaver to the Potteries and Birmingham. This alarmed the North Staffordshire Railway sufficiently for it to obtain an Act to enlarge the western end of the Trent & Mersey. There was little result, apart from some improvements to enable wide barges, after coming up the Anderton Lift, to go as far as Middlewich. Later, some struc-

6/4 Foxton inclined plane lift: the bottom basin and one of the tanks. Rails for the other tank can be seen in the foreground. The basin, and the bridge in the background, still exist; the basin is now used for pleasure craft moorings.

tures on this section were again narrowed.

In Ireland, the Lagan Navigation took over the struggling Coalisland and Ulster Canals in 1888, and the Grand Canal acquired the Barrow Navigation in 1894. Neither seems to have done particulary well out of its acquisitions.

The Grand Junction Canal Company had proposed amalgamation with other canals on the lines to Birmingham and the Trent as early as 1863, without result. In 1894, however, at the instigation of canal carrier Fellows Morton & Clayton Ltd, the GJC was able to purchase the Grand Union and the Leicestershire & Nor-

thamptonshire Union Canals, which meant that it had a continuous line as far north as Leicester. The GJC had given up carrying in 1876, and Fellows Morton & Clayton Ltd was by far the largest carrier operating on its canal; it intended, if the Grand Union were improved by dredging, and widening the narrow locks at Watford and Foxton, to operate wide boats up to the Derbyshire coalfields, which it considered would be economic where narrow boats were not. The canal-borne coal trade from Derbyshire to London, once flourishing, had almost collapsed by this date, probably a consequence of the opening of the Midland Railway's line through to

London in 1868 and the Erewash Canal's reluctance to lower tolls.

Out of this intention came the decision by the Grand Junction to install an inclined plane lift at Foxton: its tanks were to be big enough to take one wide boat or a pair of narrow boats, and raise or lower them past the two successive staircases of five locks. It was steam-powered and opened for traffic in 1900, and there is more about it in chapter seven. Too late to revive the coal traffic, too early for economical working by electricity, Foxton inclined plane was not a commercial success and was used regularly only until 1910, after which boats reverted to the locks. In the meantime the company had decided against widening the locks at Watford.

The culmination of this period of canal history came with the appointment in 1906 of the Royal Commission on Canals and Inland Navigations.

Winson Green Bridge (grid reference SP 043879) crosses Telford's new main line of the Birmingham Canal Navigations. Brick-built, but high and wide and on a skew to enable the road to cross over without diversion, this bridge and its companion in the distance mark the ultimate development of the brick-arched canal bridge. They were built in 1826. The canal is wide and straight, in contrast to the narrow and winding original line, and has dual towpaths for heavy traffic.

Lock 20 at Wyndford, Forth & Clyde Canal. The canal was closed in 1963 but here one might suppose it still in use, were it not for the fixed concrete bridge which has replaced a bascule-bridge across the lock chamber. It was from this point that pioneer steam paddle tug Charlotte Dundas *set out on her demonstration run in March 1803, successfully hauling two laden vessels over the nineteen miles to Glasgow against a headwind so strong that no other vessel on the canal attempted to move to windward. By doing so she proved the practicability of mechanical power for boats. The lock cottage, like others elsewhere, is in the vernacular style of the region.*

6/5 Cromford Wharf, Cromford Canal, in days of trade: narrow boats, piles of coal, coal merchants' carts.

6/6 (right) Part of Hatton flight, Grand Union Canal. The wide locks built in the 1930s can be seen, and alongside them the remains of former narrow locks.

The Commission investigated the inland waterways of the British Isles with great thoroughness, compared them with those of the Continent, and in 1909 issued its report in no less than twelve volumes. Principal recommendations included enlargment of canals radiating from Birmingham to London and the Rivers Severn, Weaver and Trent; and since the proposals were unlikely to be profitable enough to attract private enterprise, for nationalisation of the principal waterways.

No action resulted.

Internal combustion

The principal reason why the Royal Commission's advice went unheeded was that, by coincidence, it had come at the end of a long period of stability, immediately before a period of great change: the internal combustion engine was about to become reliable enough for widespread use, the First World War was soon to intervene.

Experiments were carried out with early motor barges and narrow boats in the 1900s: the first carriers to adopt motor craft in a big way were the Grand Canal, from 1911 onwards, and Fellows Morton & Clayton from 1912. Motor barges and boats subsequently came into general use. Development of the internal combustion engine was particularly advantageous for narrow canals, for although steam narrow boats were used, the steam plant was too bulky to be generally economic. They were common only on the Grand Junction: there, a steamer could tow a horse-boat as 'butty' and pass through the wide locks together with it. With the coming of the diesel engine, the narrow boat was freed from dependence on the horse. Yet all was not advantage, for more and more motor craft meant more and more wash and so worse and worse bank erosion, to be counteracted by bank piling with consequent expense.

The First World War had the effects not only of turning people's minds to other things than waterway improvements, but also of greatly encouraging the development of the motor lorry; and after it was over, large numbers of army-surplus motor vehicles became available. Now canal and rail alike began to feel the draught of road competition. Motor lorries carried goods from door to door without transhipment, not from wharf to wharf or siding to siding.

The effects were soon clear. The Rochdale Canal, which had had its own carrying department since the 1880s, soon gave it up. In 1921 the Leeds & Liverpool Canal gave up carrying merchandise in its own boats. The same year the Shropshire Union disbanded its large carrying fleet, after astronomic losses, caused in part by

79

Derelict chambers of the double lock (or staircase, in English terms) at Blanchardstown, Royal Canal, Ireland. Grey limestone is as typical of Irish canals as is red brick of those of the English Midlands. The canal was closed for navigation in 1961, but the top gates of the lock—on which the photographer was standing—have been restored and hold up water for local boating, all part of more ambitious plans for restoration of the whole canal. This part of the Royal Canal was built in the 1790s.

Canal age ironware: typical ground paddle gear at a lock. A windlass has to be fitted on to the main horizontal shaft to wind it, and the rack is attached to the paddle below by the vertical rod. This particular example of paddle gear, beside the Tame Valley Canal at Perry Bar, incorporates a simple safety catch engaging in the rack, reduction gears, and a cast iron post. There were many variations in detail from canal to canal, examples of which can still be seen and indeed used.

the ending of wartime subsidies. The company itself was merged fully with the LNWR as a preliminary to the amalgamation of that company itself into the London, Midland & Scottish Railway at the railway grouping of 1923.

The Stroudwater Navigation paid its last dividend in 1922, the Louth Navigation was closed in 1924, the Peak Forest Tramway was last used in 1923 and closed in 1925. The Great Western Railway, faced with mounting losses on the Kennet & Avon, attempted in 1926 to seek powers to abandon it and then, encountering opposition, offered to give the canal away free to any body considered responsible by Parliament. Since no such body appeared, the GWR was obliged to continue to maintain the canal. The Thames & Severn was less fortunate: most of it was closed in 1927 and the rest in 1933.

Traffic on the Lagan Navigation started to fall away as a result of road competition. The Grand Canal, on the other hand, in order to counteract railway road-delivery services, purchased motor lorries and set up its own road services in 1931.

Formation of the Grand Union Canal

However the most important positive step taken by canals at this period was the merger in 1929 of the Grand Junction Canal Company with all but one of the other canals on the London-to-Birmingham route: the Regent's Canal, the Warwick & Napton Canal and the Warwick & Birmingham Canal. The name Grand Union Canal was revived for the combined concern, which then purchased the Leicester and Loughborough Navigations and the Erewash Canal. The new company set about widening the canal between Braunston and Birmingham, assisted by a Government grant towards the cost. The Oxford Canal had stayed aloof from the merger, but the Grand Union was able to include its Braunston-Napton section in the scheme.

Wide locks were built, and com-

pleted in 1934, but the total plan, which included widening bridges and the bed of the canal so that wide barges could pass, was never finished, and traffic was worked by narrow boats in pairs, each motor boat towing a butty. Widening the locks did markedly reduce the time taken for pairs of boats to work through them, so value was obtained from the scheme even though incomplete. A Government grant towards widening the locks at Watford and Foxton was refused.

To promote traffic on its improved canal system the GUC Company set up the Grand Union Canal Carrying Co. This was to be the largest, and one of the last, attempts to build up and operate a fleet of narrow boats. At its greatest, the fleet included 186 pairs, most of them built new for it. This was too ambitious: the number of pairs in use never exceeded 119, in 1937. The main problem was not lack of traffic, but lack of crews. Much new traffic was gained, notably iron and steel imported through the Regent's Canal Dock, where that canal joins the Thames, and carried to Birmingham; but new crews for family boats were difficult to find by the 1930s. Meanwhile, despite its success in gaining traffic, the carrying company was running at a loss, for interest had to be paid on money borrowed to build the new boats, whether they were in use or not.

The Leeds & Liverpool Canal Co. took a comparable step to the Grand Union when in 1930 it sponsored the formation of Canal Transport Ltd, an amalgamation of four of the carrying firms on the canal with half its capital provided by the canal company. A further scheme prepared in 1920 to enlarge the Sheffield & South Yorkshire Navigation had made no progress, but some comparatively minor improvements were made in the early 1930s (lengthening of one lock, a new wharf at Doncaster).

Elsewhere, decline continued. The Ulster Canal was abandoned in 1931; so was much of the St Helens Canal. The Pocklington Canal saw its last barge in 1932, the last occasion prior to restoration when a narrow boat traded to Stratford-upon-Avon was in 1933, and the last toll was taken on the Brecon & Abergavenny Canal the same year. Traffic ceased on the Union Canal about 1933 and on the Monkland Canal in 1935. The last time there

was a working trip over the full length of the Rochdale Canal was in 1937, and that year also the Nottingham Canal was abandoned, apart from the short section in Nottingham itself which links the east end of the Beeston Cut to the River Trent; this became part of the Trent Navigation.

Two serious breaches occurred in 1936 on canals owned by the LMS Railway: in neither case were they repaired. The breach in the Bury branch of the Manchester, Bolton & Bury Canal near Prestolee must have been the more spectacular, for the canal there is carried high up on the side of the Irwell Gorge, but the breach of the Shropshire Union Canal near Frankton, on the line to Llanymynech and Newtown, had more serious consequences. For while local traffic on the Bury branch above the breach was to continue for some years, the breach on the Newtown line was made an excuse to seek powers to abandon. The sole remaining carrier was eventually compensated, though far from generously. Traffic died out over the rest of the one-time Ellesmere Canal in 1939. The same year it ceased on the pioneer Newry Canal, and the Duke of Bridgewater's flight of locks at Runcorn was closed, later to be filled in.

The Second World War and its aftermath

The Second World War brought about a mild resurgence of traffic on some canals. Those British canals and river navigations still considered important for transport were taken under Government control, as were almost all railways. This meant that railway-owned canals were controlled too, but by 1939 many of these were in very bad condition or un-navigable, despite legislation to the contrary. The LMS obtained powers to close part of the Manchester, Bolton & Bury in 1941; then in 1944 it obtained an Act of Parliament authorising closure to navigation of no less than 175 miles of its canals, all or nearly all of which were no longer used for trade. Hardest hit was the Shropshire Union system, for closed under the Act were the whole of

6/7 Leeds & Liverpool Canal short boats at Blackburn in the 1940s.

the former Montgomeryshire, Ellesmere and Shrewsbury Canals (except the Ellesmere's Wirral Line), what remained of the Shropshire Canal, and the Newport branch. Other canals under this Act were the Huddersfield Canal, most of the Cromford Canal, the Leek branch of the Trent & Mersey, and the top end of the Ashby.

The darkest hour is that before dawn. In that same year there was published the book *Narrow Boat* by the late L. T. C. Rolt: his first book, one which gained instant popularity, and the origin of present-day interest in canals.

In his book Rolt described a voyage made in 1939 and 1940 in the narrow boat *Cressy*. Originally a Shropshire Union fly boat, *Cressy* had been converted into a mobile houseboat fitted

with a Ford Model T engine. The voyage described in the book took her from Banbury to Leicester and up the Trent & Mersey as far as Middlewich, then south again by the Shropshire Union, Staffs & Worcs, Trent & Mersey, Coventry and Oxford Canals. People had cruised canals and written about their experiences before— among them were J. B. Dashwood, who traversed the Wey & Arun just before it closed in 1868, P. Bonthron, whose book *My Holidays on Inland Waterways* is a record of 2,000 miles of pleasure cruising in the period before the First World War (it gives a fascinating insight into the condition of canals at that time), and Temple Thurston, who gave an attractive picture of a voyage by horse-drawn narrow boat along the canals and

rivers of the South Midlands in *The 'Flower of Gloster'* first published in 1911. But none of these had the impact or consequences of *Narrow Boat*. In his book Rolt vividly described an England he had found which was scarcely known to the general public, one in which little had altered for a hundred years. *Narrow Boat* was an immediate success.

Among those who read it was Robert Aickman, whose attention had already been drawn to canals by the near-derelict state of the canal at Stratford-upon-Avon. He corresponded with Rolt and in 1946 the Inland Waterways Association was formed, with Aickman as chairman and Rolt as secretary, its principal object being to advocate the use, maintenance and development of inland waterways in the British Isles. The Inland Waterways Association has become one of the most influential of pressure groups; its formation was followed in due course by the Inland Waterways Association of Ireland, the Scottish Inland Waterways Association, and numerous local waterway societies, large and small.

Another consequence of *Narrow Boat* was to stimulate greatly pleasure traffic on the canals. There had always been a little pleasure traffic, some instances of which have already been mentioned. In addition to these, during the 1880s horse-drawn pleasure boats had started to operate on the Shropshire Union Canal at Llangollen, as their successors continue to do, and towards the end of the nineteenth century the great interest in rowing and canoeing resulted in publication of guides to inland waterways for oarsmen. Between the wars there were certainly a few other narrow boats converted, like *Cressy*, into mobile houseboats, and motor cruisers for canal holidays were first hired out at Chester in 1935.

Nevertheless pleasure cruising was rare on canals, and it was only after publication of *Narrow Boat* that it started to grow. It developed in three principal modes: by privately-owned boats, by hire craft, and by hotel boats—that is, narrow boats converted into travelling hotels. From that time onwards, pleasure cruising, the IWA, and interest in and enthusiasm for canals have grown continuously, mutually encouraging and reinforcing one another.

On 1 January 1948 came nationalisation of transport in Britain, under the Transport Act 1947. It covered all the railways and waterways which had been under Government control during the Second World War; the list of waterways included:

Aire & Calder Navigation
Sheffield & South Yorkshire
   Navigation
Birmingham Canal Navigations
Calder & Hebble Navigation
Coventry Canal
Grand Union Canal
Leeds & Liverpool Canal
Oxford Canal
Trent Navigation
Severn Navigation
Gloucester & Berkeley Canal
Worcester & Birmingham Canal
Staffordshire & Worcestershire
   Canal
Stourbridge Canal
Weaver Navigation.

Also nationalised were the canals owned by railway companies. The most important waterway not to be nationalised was the Manchester Ship Canal, considered to be a port, and its satellite the Bridgewater Canal. A few waterways no longer considered important for transport were also omitted from nationalisation; these included the Basingstoke Canal (on which the last freight traffic passed in 1949), the Rochdale Canal, the Stroudwater Canal and some river navigations.

The undertakings of the nationalised companies were vested in the British Transport Commission; this operated through a series of public authorities called executives. The canals which had been independent passed directly to the Docks & Inland Waterways Executive and the railway-owned canals were transferred to it gradually from the Railway Executive. Generally, nationalised canals (unlike railways) continued to be described by their old names—Grand Union, Oxford, Leeds & Liverpool and so on. There have, however, been instances where a name has altered, over the years, from the original. The Gloucester & Berkeley is now known as the Gloucester & Sharpness, and the canal from Hurleston Junction to Llangollen and Llantysilio, originally part of the Ellesmere Canal and then of the Shropshire Union, is often referred to as the Llangollen Canal. The Caldon branch of the Trent & Mersey is often called the Caldon Canal.

Despite the decline of traffic which had taken place, some of the old-established canal companies were paying dividends right up to nationalisation. The Coventry and the Staffs & Worcs were among them. The Grand Union, too, had been getting its finances straight—and this included the carrying company, which had sold off many of its surplus boats. Independent carriers were not nationalised, though Fellows, Morton & Clayton Ltd went into voluntary liquidation in 1948 and its fleet was acquired by the British Transport Commission.

In 1950 the Grand Canal was merged with Coras Iompair Eireann, the state undertaking responsible for public transport in the Irish Republic. The Royal Canal, previously railway-owned, already belonged to it; the last traffic moved on this canal in 1951. Canals in Northern Ireland were not

nationalised but little traffic remained. The Newry Canal was abandoned, most of it, in 1949.

In Britain, the decline went on. In 1950 the Monkland Canal was abandoned, in 1951 the last traffic passed through Dudley Tunnel, in 1952 the Rochdale Canal was closed except for the short section in Manchester which gave access to the Ashton Canal and, through it, the Peak Forest Canal. Then in 1954 traffic ceased on these canals also. The same year the boatage services in the Birmingham area were withdrawn, and the Stroudwater Canal was abandoned (not without opposition from the IWA) except for a short section at Saul, its junction with the Gloucester & Sharpness, which is still used as moorings. The Coalisland Canal and most of the Lagan Navigation were also abandoned in 1954, the remainder of the latter in 1958. In 1955 the BTC obtained the authority of Parliament to close the long disused Ripon Canal, and the northern part of the Lancaster Canal, on which trade had last moved in 1947.

With the establishment of British Transport Waterways at the beginning of 1955, administration of inland waterways was separated from that of docks. A BTC survey later that year recommended that, of its waterways, 336 miles should be developed, 994 miles retained, and 771 miles, which were either disused or carrying too little traffic to justify retention for commercial navigation, should cease to be its reponsibility. In other words, it seemed, the 771 miles were proposed for closure.

The pro-canal faction was now growing in strength, and when the BTC promoted a Bill in Parliament to authorise abandonment of the Kennet & Avon from Reading to Bath (already partially and illegally unnavigable), a petition to the Queen against closure was signed by about 20,000 persons living near the canal.

As a result of controversy aroused by the survey report and the K & A proposal, the Government set up an independent committee of enquiry into the future of inland waterways. This committee, called the Bowes Committee after its chairman, reported in 1958 and accepted the viewpoint put forward by the IWA that canals should not be considered solely in terms of transport, but rather that they had a multiplicity of existing uses—water supply (to factories and farms), land drainage, boating, fishing, amenity. This was in turn accepted by the Government.

The downhill trend in traffic continued, with occasional reactions. Regular traffic on the Shropshire Union through Ellesmere Port ceased about 1958, but in 1959 a new long lock at Long Sandall on the Sheffield & South Yorkshire enabled trains of compartment boats to reach Doncaster.

CIE ceased to carry by barge in 1960; bye traders, or independent carriers, had ceased to operate about 1957. Since 1960 the principal lines of the Grand Canal, from the Liffey at Dublin to the Shannon and the Barrow, have been kept open and used by pleasure traffic; various branches have been closed. The Royal Canal was closed to navigation in 1961.

Restoration of derelict canals

Many canals lingered on after traffic ceased, the condition of their locks deteriorating and their channels becoming overgrown, silted and impassable. Abandonment of a navigation, when it came, did not eliminate a canal physically. It went on being a liability to its owners, with bridges to be maintained and breaches to be avoided. Even to empty a closed canal was not straightforward, for canals had become part of the land drainage system of the country through which they passed, and furthermore they were often affected by long term agreements or obligations towards water supply.

By 1960 the IWA had formed the opinion that canals which were derelict or nearly so could be restored for navigation more cheaply than the BTC suggested if voluntary labour and support were used. As early as 1950 the Lower Avon Navigation Trust Ltd had been formed, under the wing of the IWA, to restore the nearly derelict Lower Avon (river) Navigation from Tewkesbury to Evesham, using such methods, and was proving to be a success; and in 1959 a short length of derelict canal, the Wyken Arm off the northern Oxford Canal, had been dredged and restored by volunteers. In

1960 the IWA had the opportunity to put its ideas fully into practice on the Stratford Canal.

The southern part of this canal, from Lapworth to Stratford-upon-Avon, was by then impassable to anything larger than a canoe. It was still a statutory navigation, but abandonment was proposed. After long negotiations, it was transferred from the BTC to the National Trust on a five-year lease with an option, which was eventually exercised, for the National Trust to acquire the canal permanently. Over the next three and a half years the canal was restored, largely by voluntary labour, and reopened by the Queen Mother in 1964. No regular freight traffic reappeared, but pleasure craft certainly did. Restoration of the Lower Avon and the Stratford Canal was the start of a nationwide movement towards restoration of disused and un-navigable waterways by similar means: the story is dealt with in more detail in my book *Waterways Restored*.

Elsewhere in 1961 the British Transport Commission started work on a modernised freight depot at Rotherham, on the SSYN, in anticipation of enlargement of that navigation. The Dearne & Dove Canal, however, which connected with it at Swinton, was closed except for a short length at the junction. The remaining sections of the Manchester, Bolton & Bury were closed also.

British Waterways Board established

Nationalised transport was extensively reorganised as a consequence of the Transport Act 1962. The BTC was abolished and its waterways transferred to a new public authority, the British Waterways Board. The BTC had been rail-oriented, but the new BWB was responsible solely for waterways. It was to provide services on its inland waterways so far as it thought expedient, and review how those waterways not required for the purpose might be put to best use.

The winter of 1962–63 was extremely severe, and many canals were icebound for long periods with consequent diversion of traffic to other forms of transport. Following this, during 1963 and 1964 British Waterways Board ceased to carry by narrow boat (with the noted exception of the lime juice run from Brentford to Boxmoor on the Grand Union Canal which it retained for some years), and also ceased carrying on the Leeds & Liverpool Canal where it had inherited Canal Transport Ltd's fleet. Independent carriers continued to operate on the canals affected, Willow Wren being notable among them.

Cargo traffic had ceased on the Forth & Clyde Canal about 1957, but it was still used by fishing vessels and pleasure craft crossing from one coast to the other. Intended for seagoing vessels, it had been laid out with unlimited headroom for masts: all the bridges opened. So by the 1960s passage of a boat meant extensive delays to road traffic, and finally in order to avoid the cost and inconvenience of an opening bridge on a new main road, the A80, the canal was closed to navigation at the beginning of 1963. The St Helens Canal was abandoned the same year and the Union Canal was closed two years later. In 1966 commercial traffic ceased at the western end of the Bridgewater Canal and the remaining flight of locks connecting it to the Manchester Ship Canal at Runcorn was closed.

By 1965 the BWB had completed a detailed review of its waterways, and published the results in book form as *The Facts about the Waterways*. There were two principal conclusions. The first was that a very small part of the Board's waterway system was still viable commercially—mostly the waterways radiating from the Humber, together with the Weaver and the Severn. All small canals were excluded. For these, a waterway by waterway analysis showed that if all were treated in the cheapest possible way, whether by elimination or conversion into unnavigable water channels, the Exchequer would still have an inescapable bill equivalent to £600,000 a year (1965 prices). To keep such non-commercial waterways open for pleasure craft was estimated to cost an additional £300,000 to £350,000 a year.

These findings were reflected in the

Transport Act 1968 which classified the BWB's waterways in three categories:

Commercial Waterways, about 300 miles, to be principally available for the commercial carriage of freight.

Cruising Waterways, about 1,100 miles, to be principally available for cruising, fishing and recreation.

The remainder, about 600 miles, which were to be treated in the most economical manner possible, consistent, where they were retained, with public health, amenity and safety.

The principal commercial waterways were, and are, those which radiate from the Humber, and the Severn, the Gloucester & Sharpness Canal, the Weaver, the Caledonian Canal and the Crinan Canal. The cruising waterways were established as all the main through routes of the nationalised canal system which were then navigable, and some of their branches. A notable inclusion was the Llangollen Canal. Abandoned in 1944, it had been retained for water supply, and had become extremely popular for pleasure cruising.

Of the remainder waterways, about 350 miles had already been statutorily closed for navigation, and much but not all of the remaining 250 miles was no longer navigable.

The Act also established the Inland Waterways Amenity Advisory Council, to advise the Board and the Secretary of State for the Environment.

All this was, in the circumstances, useful and generally constructive, but there was one provision in the Act which was the opposite. This was a general abolition of a public right of navigation over the Board's artificial waterways, coupled with repeal of section 17 of the Regulation of Railways Act 1873 mentioned above. The Board was given a duty to maintain the commercial and cruising waterways to proper standards, but the public appears to have no right to navigate them.

6/8 (preceding page) Pleasure
and commercial use overlap: a
yacht passes a Clyde Puffer at
Ardrishaig, Crinan Canal,
during the early 1950s.

6/9 (left) A pair of laden Samuel
Barlow Coal Co. Ltd narrow
boats on the Grand Union Canal
near Bugbrooke, Northants,
during the last years of trade.

quired canal-borne coal; and the last regular coal traffic on the Leeds & Liverpool (which was carrying 800,000 tons a year as recently as 1958) ceased in 1972 when a power station at Wigan changed its source of supply.

On the Bridgewater canal, regular commercial traffic ceased in 1974. The last goods carried had been imported corn, from Manchester Docks back along the Duke's original canal to Kellogg's Mill near Barton Aqueduct. After Kellogg reorganised, it was no longer required.

When commercial traffic ended, it seemed to the Manchester Ship Canal Company that it should not carry the whole of the operating loss on the Bridgewater Canal, and so a trust was established: its trustees are the company and the riparian local authorities. This trust supervises the company's running of the Bridgewater Canal, and shares in the financial burden.

The present day

Day boats still convey refuse around Birmingham, and since many narrow boats have passed into the hands of enthusiasts, who use them when the opportunity arises, the sight of long-distance narrow boats in use is not lost. An enterprising development has been the retail sale of coal from canal boats. This originated in 1973 with the desire of enthusiasts on the Ashby Canal not only to perpetuate narrow boat carrying but also to help keep the canal clear of weed and obstructions by regular passage of loaded boats. They established Ashby Canal Transport Ltd; one of its activities is that of coal merchant, and the coal is carried by boat for sale to canal-side premises. On the Leeds & Liverpool, Apollo Canal Carriers has successfully revived some grain traffic.

Continuing traffics on the large

Traffic yesterday and today

The years immediately following the 1968 Act were marked by development of improved commercial craft in York- shire, notably the modern compart- ment boats on the Aire & Calder where a push-tug and three boats convey 480 tons of coal at a time, and by the end, with very few exceptions, of regular commercial carrying on small canals elsewhere.

It is worth pausing for a moment to recall what cargoes were being carried in the last years of regular trade on the small canals. Coal, of course, and timber and grain were important. Sugar and wool and tar were carried too, and there were unlikely-sounding traffics such as aluminium ingots, tomato purée, and chocolate crumb.

The last traffic on the Trent & Mersey Canal to the Potteries ceased about 1969; the last long-distance coal traffic over the northern Oxford and Grand Union Canals in 1970 when a Southall jam factory no longer re-

waterways of Yorkshire include coal, oil, timber, grain, steel and sand. The Aire & Calder has been steadily improved: in the 1950s the largest craft which could use it had a capacity of 250 tons, but the locks have been lengthened and curves eased so that from 1978 vessels carrying as much as 700 tons can reach Leeds from the Humber. The long-proposed enlargement of the Sheffield & South Yorkshire Navigation has fared less well: British Waterways Board put up a scheme to the Government in 1966, was turned down, and tried again in 1972. Since then SSYN improvement has been both political shuttlecock and cause célèbre. At the time of writing the Government, after years of resistance, has suddenly given its approval: though where the money is to come from is not yet clear.

Canal pleasure traffic has grown enormously. At the end of the Second World War it was virtually non-existent; in 1951 the BTC issued 3,700 permits for pleasure craft, in 1968 (the year of the Transport Act which retained much of the canal system for pleasure cruising) BWB issued a total of 12,888 craft licences and lock passes, and in 1977 the total number of pleasure craft licensed by BWB was 23,001.

The waterway restoration movement has made great progress also. The Waterway Recovery Group of the IWA has evolved into a national co-ordinating body for voluntary work, and many local authorities have contributed financially, particularly towards work on remainder waterways. So, many canals which were unnavigable, or almost so, have been restored: these include the Stourbridge Canal, the Caldon branch to Froghall, the Ashton and Peak Forest Canals between Manchester and Marple, and the Brecon & Abergavenny Canal. Two isolated lengths of BWB Canal have been transferred to local authorities: the Grand Western and the northern part of the Cromford. Both

have since been restored as amenities with voluntary support. Much of the Kennet & Avon Canal and about half of the Pocklington Canal have been reopened for navigation and work on these continues; elsewhere, useful work is being done on (among others) the Basingstoke, Montgomery and Thames & Severn Canals.

The position with maintenance of the commercial and cruising waterways since 1968 has been less satisfactory, for although Parliament gave BWB a mandate to maintain them, the Government has starved it of finance. This was not immediately obvious: it

emerged gradually, to be confirmed by the Fourth Report of the House of Commons Select Committee on Nationalised Industries, published in 1978.

The Board's sources of finance are grants and loans from the Department of the Environment, Treasury-guaranteed bank overdrafts, and trading income; year by year a grant-in-aid has been made by the Department.

In 1970 the Board advised the DoE that a big engineering programme to cost £21·8 million was needed to overtake arrears of maintenance arising from under-expenditure in past

6/10 The last of Winwick Lock, St Helens Canal, during infilling operations, April 1978. It was, perhaps, being preserved for a future generation of archaeologists – provided they do not mind what they dig through!

years (i.e., before 1968). According to the select committee report the Department 'did nothing for four years', and then it appointed independent consultants, Peter Fraenkel & Partners, to survey the system. In due course the Board's own findings were confirmed, but although the Fraenkel Report was submitted to the DoE in January 1976, it was not published until November 1977. At the same time the Government allocated £5 million in 1978–9 for urgent maintenance involving public safety; a further £20 million has since been promised over five years.

By 1977 the cost of works needed had inflated to about £60 million, and the effects of inadequate maintenance have become all too obvious: frequent, sudden and lengthy stoppages for repairs to dangerous structures such as aqueducts and tunnels, and canals with eroded banks and silted channels, for where canal banks are unprotected, they are eroded by the wash of passing boats and the resulting silt is deposited in the channel.

In saying that the DoE did nothing for four years from 1970 the select committee was less than correct. What it did do, in 1971, was to make

proposals for dismemberment of the BWB system and its distribution among regional water authorities. This would have had the effect of passing down the costs of maintaining canals from national to local government, from taxpayer to ratepayer. These proposals provoked much opposition and were defeated.

In 1977, however, came a further proposal of the same sort. According to the White Paper *The Water Industry in England and Wales: the Next Steps,* BWB should be merged with a National Water Authority and the water industry would assume major financial responsibility for the waterways. The expenditure necessary to meet the maintenance backlog made it 'essential to take a new look at methods of making the system viable . . . considering for individual waterways the cost of maintaining them in relation to the benefits they provide to transport, recreation and amenity, and the needs of the water industry.' Yet another survey, when what is needed is action.

On to this scene came the House of Commons Select Committee. It investigated the British Waterways Board during 1977, and in its report, published early in 1978, was highly critical of the Department of the Environment and the minister responsible, the Rt Hon Denis Howell MP. The Government, it recommended, should abandon proposals to merge BWB with the National Water Authority; it should accept the Fraenkel Report and finance the maintenance backlog; it should immediately approve the Sheffield & South Yorkshire enlargment scheme; this and much more.

Those who have followed canal politics were not surprised to find that in due course the Government rejected the findings of the select committee. Then, suddenly, it went back on itself and authorised the SSYN scheme. The dust has yet to settle*: before this is published there may well be further developments. Whatever they may be, it is to be hoped that adequate finance will be made available, for repairs rather than reorganisation: otherwise the future of British canals and their historic features will become most insecure.

*The Minister for Local Government and Environmental Services, the Rt. Hon. Tom King M.P., stated in 1979 BWB will remain an independent entity.

CHAPTER 7

FEATURES OF CANALS: THE WATERWAY

Contour canals

The most striking feature of early canals in England is the circuitousness of some of their routes. They slavishly follow the contours and wind across the surface of the land; cuttings and embankments, which would enable them to follow a more direct line, are almost unknown on them. The summit pound of the Oxford Canal is the most remarkable example to survive: it is 11 miles long, from Marston Doles to Claydon Top Lock, a distance of $4\frac{3}{4}$ miles as the crow flies. The original lines of the Birmingham Canal and the northern part of the Oxford were similarly winding; they were shortened later in the canal era, but traces of the original lines remain in the forms of short branches or loops off the present lines.

These were Brindley canals, though not among the earliest for which he was responsible. The lines of the Trent & Mersey and the Staffordshire & Worcestershire are comparatively straight, which seems to have been due, as much as anything, to the good fortune that they could follow straightish valleys for much of their courses, through localities which were not particularly undulating. Certainly cuttings and embankments are generally conspicuous by their absence from their routes and these canals tend to follow the surface of the ground as much as possible. In doing so the summit pound of the Staffs & Worcs does, between Four Ashes and Slade Heath, traverse three sides of a shallow re-entrant that later canals would have crossed by an embankment, and in consequence the distance by canal is $2\frac{3}{4}$ miles between points less than 1 mile apart in a straight line. The layout of this particular section of canal may perhaps be considered a foretaste of what was soon to follow!

It is interesting to speculate on the origins of the contour canal, or rather the canals which, laid out by Brindley and his pupils, followed the contours to an exaggerated extent. Charles Hadfield in *Canals of the East Midlands* quotes the Oxford Canal Company as stating in 1829 that 'in consequence of opposition on the part of the Land Owners . . . and from the art of making canals not being so well understood then as it now is . . . the canal in its Northern part was made in a direction too circuitous.' (And doubtless its southern part also, but then they were not proposing to shorten that.)

That bit about the art of making canals not being well understood may well have been true so far as cuttings, embankments and the builders of the Oxford Canal were concerned, but the technology required certainly existed elsewhere. The first sections of the Grand Canal in Ireland were cut about twelve years earlier, between 1756 and 1761, and there was no nonsense about following the contours. This canal runs straight, striding across country on long embankments and slicing through high ground by long rock cuttings.

I am inclined to attribute Brindley's liking for contour-following to his earlier experience as a millwright, constructing watercourses for mills. A mill leat, as I understand it, would parallel, more or less, the course of the stream which had fed it, following the contour until it reached its mill. So long as it delivered an adequate supply of water, its length was irrelevant. There was no need to take a direct line, no opportunity for a builder of such watercourses to find out by making embankments and cuttings. This is born out, I suggest, by construction of the embankment, associated with Barton Aqueduct, out of masonry rather than as a straightforward earth embankment. What Brindley's work on mills had brought him into contact with was mines and mining techniques (through his application of water power to mine drainage), and this would have enabled him to find out about making tunnels, which he preferred to cuttings.

And yet the Stretford to Runcorn line of the Bridgewater Canal, engineered by Brindley, is not entirely a contour canal. It has long and substantial embankments across the valleys of the Rivers Mersey and Bollin. There are also many smaller embankments, where roads (themselves sometimes in cuttings) are carried beneath the canal by 'underbridges'. Brindley the innovator, nick-named *The Schemer*, seems to have been at work, for this technique does not generally appear again until much later in the canal era. I am inclined to attribute the Runcorn line's advanced engineering—compared with contemporary canals—to personal involvement by Brindley. On other canals, with which he was associated, his title was surveyor-general or some such, with subordinates in charge of engineering detail. On the Runcorn line, the Bollin aqueduct and embankment were completed in 1767, and the line was open to Lymm in 1769 and Stockton Heath in 1772. Construction of the Trent & Mersey started in 1766, and of two notable contour canals, the Birmingham and the Oxford, in 1768 and 1769 respectively. This would not leave much time to establish for certain the stability of the earthworks on the Runcorn line. My opinion is that Brindley as a wise man would have kept the new and risky developments to himself, and passed on tried and trusted techniques to his pupils. Had he not died in 1772, he might perhaps have passed on techniques of cutting and embankment construction also.

That is conjecture. Another possible factor is that a canal which precisely follows a contour can easily intercept streams which cross its route and so gain a supply of water. In practice, however, this does not seem to have been done in many places. In any

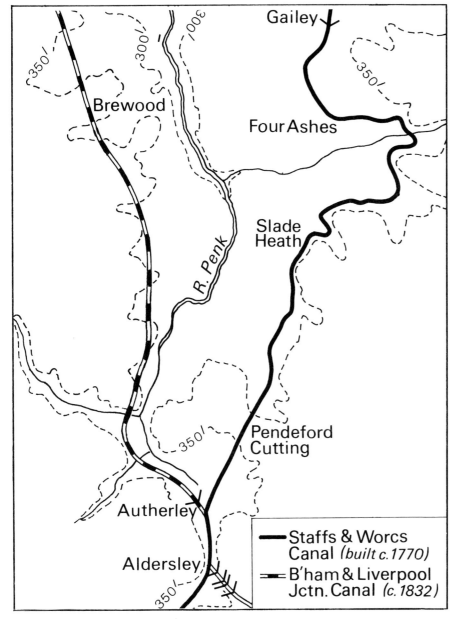

7/1 The Staffordshire & Worcestershire Canal, the Birmingham & Liverpool Junction, and the 350 feet contour. Brindley laid out the S & W's summit pound south of Gailey at 341 feet above sea level; Telford's Birmingham & Liverpool Junction is lower by the six-inch fall of the stop lock at Autherley. Nevertheless the 350 feet contour cuts across the B & LJ in many places, as the canal follows its direct course by means of cuttings and embankments. The earlier canal winds along the surface of the ground, forsaking it only to cross a tributary of the River Penk by an aqueduct, and for one cutting at Pendeford.

the sides of the Weaver valley along slopes so steep that they were obliged, in places, to tunnel behind them. Other examples of canals built along steep hillsides were the Ellesmere Canal's branch to Llangollen, the Manchester, Bolton & Bury above the Irwell Gorge near Prestolee, and several South Wales canals, notably the Brecon & Abergavenny between Govilon and Goytre.

Cuttings and embankments

Cuttings and embankments on the early canals in England were considered in isolation from one another when the canal was surveyed. Even Brindley's canals passed through occasional cuttings: the one by which the summit pound of the Staffordshire & Worcestershire Canal passes through high ground near Wolverhampton is not unimpressive (grid reference SJ 907034). Although only about 10 feet deep, it is some 600 yards long, cut through sandstone: wide enough only for a single boat, it has passing places at intervals. A shorter but much deeper and wider cutting through sandstone, dating from the same period, is that by which the Chester Canal threads the city of its name: its construction must have helped considerably to strain the finances of that unfortunate company, and it is probably symptomatic of the limited knowledge of canal building at the time that no cheaper route to the Dee was found.

Thomas Omer, laying out the Grand Canal between Dublin and Sallins in the 1750s, clearly understood the principal of cut-and-fill, that is to say, using material excavated from cuttings to form embankments. As mentioned, the canal follows a straight course, alternately along embankments and through cuttings (and, at one point, deliberately, through a limestone quarry from which material for bridge and lock construction was obtained). Omer was later criticised by Smeaton for adding to the difficulties of construction.

One wonders what Smeaton would have said of the Royal Canal's cutting at Clonsilla, outside Dublin, built in the 1790s. Here for 2 miles the canal passes through a limestone cutting

event, the rolling country of the Midlands gave rise to several other canals with circuitous contour-hugging courses, parts of the Dudley, Ashby and old Grand Union Canals being among them.

While some contour canals are circuitous in a horizontal plane, there are others of which the locations appear almost precipitous in a vertical one. In order to maintain its level, the Birming-

ham & Fazeley Canal is carried through Hopwas Woods on a ledge above the River Tame, a position which may well have contributed to the reluctance of the Coventry Canal company to build this part of its authorised line. The builders of the Trent & Mersey, having been inveigled into joining the Duke's canal rather than the River Weaver, found themselves constructing their canal high on

94

which is in places 27 feet deep: much of the cutting is too narrow for boats to pass, or even for a towpath, which is carried high above the water.

Cut-and-fill technique was adopted gradually in Britain. An example dating from the late 1780s can be seen on the Birmingham & Fazeley Canal at Hademore, between Whittington Brook and Hopwas. Here what is generally a contour canal passes through a deepish cutting. This is followed by a straight embankment some 200 yards long across a valley which, though not very deep, is extensive enough to add considerably to the length of a line which rigidly follows the contours. Much of the Grand Junction Canal shows similar characteristics.

On the canals of the 1830s, and particularly on the Birmingham & Liverpool Junction, cut-and-fill was used throughout much of the line, which was made as direct as possible, and contours were followed only where convenient. The method was adopted soon afterwards for railways and the line of the B & LJ has more in common with a railway than an early canal. It has some notable cuttings— particularly Woodseaves (grid reference SJ 697306). This cutting is $1\frac{1}{4}$ miles long and as much as 70 feet deep, cut through alternating strata of friable rock and clay which gave much trouble with slipping when the cutting was new and continue to slip to this day. It is impressive: more interesting and unusual are the cuttings at the lock flights at Tyrley and Audlem. On other canals, locks were placed, at ground level, in positions where the slope of the surface was suitable, but on the B & LJ cut-and-fill continues down flights of locks. The canal has to cross some narrow valleys during the course of these, and directness of line, smoothness of gradient and equality of cut-and-fill are maintained by placing many locks in the cuttings which alternate with embankments.

The most interesting of all canal cuttings is Galton or Smethwick Cutting, near Birmingham. From Brasshouse Lane Bridge (grid reference SP 019888) three levels of canal can be seen, representing the work of three of the greatest canal engineers. Deep in the bottom of the cutting is Telford's broad new BCN main line, running straight as a die; higher up on a ledge on the north side of the cutting is the old main line, in fact Smeaton's work, still in water and in use; and higher up a narrow and dry ledge marks the site of Brindley's original summit pound.

Embankments, like cuttings, on early canals in England were generally avoided and, where this was not possible, considered in isolation. The Staffordshire & Worcestershire, for instance, could not do other than approach the Trent & Mersey at Great Haywood by an embankment: traces of excavations alongside it suggest that it was built up of material excavated from them.

The middle years of canal construction produced some notable embankments, such as the 97-feet high southern approach embankment to Pontcysyllte Aqueduct and the embankment, $\frac{3}{4}$ mile long, which carries the Leeds & Liverpool nearly 60 feet above Burnley. The embankments, which continue for several miles, by which the Grand Canal crosses the Bog of Allen are that canal's greatest engineering feature: they took many years to construct for, as the canal was built, the bog in its vicinity was drained and its level subsided, so that what had hopefully been intended as a canal at or near ground level eventually became carried on a continuous high embankment. The 4-mile stretch between 20th Lock and the junction with the Edenderry branch gave as much trouble as any, and is today as typical.

Shelmore Embankment on the Birmingham & Liverpool Junction (grid reference SJ 796220) has the most epic story of all. A mile long and 60 feet high, it took 6 years to build, so badly did the marl out of which it was built tend to slip. Today it gives holiday boaters an impressive view over the Staffordshire countryside: yet it was necessary only because of the obdurate attitude of a landowner who would not permit the canal to be made at ground level through his pheasant coverts nearby.

A breach in such a location would be

7/2 (right) Pleasure craft approach Braunston by the Oxford Canal along the high and straight embankment built as part of the improvement scheme of the early 1830s (grid reference SP 531659).

serious, and the B & LJ and similar late canals such as the Tame Valley have stop gates at strategic locations—gates similar to lock gates, which can be quickly closed across narrows in the canal in emergency. The older method, still in use on other canals, is to use stop planks, inserted one on top of another in grooves built into walls at the sides of the canal. These are generally at bridge holes where the canal narrows in any event, and stop planks stored under cover nearby are a common sight.

Elsewhere canals expand into winding holes where full-length boats may be winded or turned round, for they are usually longer than the width of the canal.

The banks and the towpath

Canal beds were and are made watertight where necessary with puddled clay, and their banks were originally left generally unprotected, for unpowered boats made little wash. Powered craft produce wash waves which damage banks and, since the material washed away settles on the bed of the canal, increase the need for dredging. The Bridgewater Navigation Company largely replaced horse haulage by steam tugs on the Bridgewater Canal in the late 1870s, and in consequence had to line the banks throughout with stone walling. This remains and today provides useful protection against the wash of pleasure craft. The Grand Union, during its improvement schemes of the 1930s, lined banks with concrete piles, a method which was continued by the nationalised waterways until the late 1960s when BWB went over to steel piling which is cheaper but shorter-lived.

Bank protection work needed to make BWB canals wholly suitable for power craft has never been completed and the Fraenkel Report estimated in 1975 that, although 75 per cent of towpath banks and 40 per cent of offside banks did have some form of protection, of this more than a quarter needed repair and just less than a quarter needed replacement; furthermore all unprotected banks on the towpath side and about half of those on the offside needed protection. Cer-

tainly it has become very noticeable in the past few years that the banks of the most popular cruising canals have, where unprotected, become very badly eroded, even, on the Shropshire Union for instance, to the extent that the towpath is in places completely washed away. These canals have in turn become very shallow. It is all a consequence of rapidly increasing use of an inadequately financed canal system.

Of all surviving traces of canals' horse-powered past, the towpath is the most prominent. Generally it was placed on the downhill side of a contour canal; when the canal was built, the excavated material was dumped on this side to retain the water and support the path. A hedgerow or fence accompanied the towpath to prevent animals straying and mark the boundary of the canal company's property. On the offside of the canal, the water's edge was generally the boundary: where additional land—a cutting side, say—was owned by the canal, the boundary was indicated by a row of boundary markers. These were usually small posts of earthenware or cast iron, bearing the company's name or initials. They can be seen alongside the Coventry, Ashby and Grand Junction Canals.

7/3 Milestone in the towpath hedgerow at Gayton Junction, Grand Union Canal, is headed with the initials of the Grand Junction Canal Co., original owner of this section. It indicates distances to its junctions with other waterways.

That applies in the country. In towns, canals are often hemmed in by walls, or the backs of factories and other buildings. Town canals are today often fascinating but seldom ornamental. Notable exceptions are those few places where advantage was taken, in the canal era, of a canal's potential to embellish a town or city. Examples are the Regent's Canal at Regent's Park and Little Venice, London, the Grand Canal Circular Line in Dublin, and the Kennet & Avon through Sydney Gardens, Bath. In the country a comparably ornamental location is Tixall Wide, on the Staffs & Worcs near Great Haywood, where the canal broadened into an ornamental lake so that it might embellish the view from a landowner's residence.

Along the towpaths were and are milestones, formerly essential for calculation of tolls. They are sometimes literally of stone, on the Manchester, Bolton & Bury, Royal and Union Canals for instance, and sometimes of cast iron (eg: Grand Junction, Shropshire Union, Trent & Mersey). Study of their inscriptions is sometimes revealing about canal history. Those along the Birmingham & Liverpool Junction indicate the distances not only to the ends of the canal at Autherley and Nantwich, but also to intermediate Norbury, where once the branch to Newport and the Shropshire canals joined the main line; and at Etruria, where the Caldon branch of the Trent & Mersey meets its main line, a milestone still shows the distance to Uttoxeter, although the canal beyond Froghall was closed as long ago as 1847.

Towing-paths were generally surfaced, if at all, with gravel, and only where width was restricted, beneath a bridge for instance, were they paved. In 1978 I found the Black Country Museum reproducing a paved towpath in such a situation, with traditional materials: blue brick pavers with crisscross recesses in the familiar diagonal pattern, and ridges of brick scorchers laid on edge across the path for horses' hooves to grip. The towpath bank was edged with blue brick copings and protected from impact of passing boats by cast iron bumping irons. All these features are still to be seen in many places around the English canal network, though often they are battered or weathered.

Stone sleeper blocks from super-

7/4 A stone sleeper block from a tramroad is used as a coping stone on a wharf at Gnosall, Shropshire Union Canal. The recess for the chair can be seen, and one of the deep holes, provided for the pegs which located the chair, has been used to secure a mooring ring.

seded tramroads were often re-used as copings for towpaths or wharves. Roughly cubical, and between one and two feet in each dimensions, they can be recognised by a rectangular recess, in which the cast iron chair which supported tramplate or rail was located, with two deeper round holes into which the pegs which secured it were inserted. I have noted such blocks at Hinckley on the Ashby Canal, Gnosall on the Shropshire Union and Hey Lock (or its filled-in remains) on the St Helens Canal (grid reference SJ 582938).

Tunnels

Structures built early and late in the canal era can be seen side-by-side at Harecastle, for the Trent & Mersey company did not accept Telford's advice to modernise Brindley's tunnel, and its portals remain near those of the later one: indeed, the east portal of Brindley's tunnel (grid reference SJ

849517) comes into view from an approaching boat long before that of the later tunnel through which the boat is to pass, for a long straight section of canal leads to the tunnel and this is aligned, of course, on the old tunnel rather than the new.

After Telford's new tunnel was completed, both tunnels were used for many years: eastbound boats passed through Brindley's tunnel and westbound through Telford's. Brindley's tunnel had a bore of minimal size and boats were propelled by legging—that is to say, by two men who lay head to head across the foredeck and pushed against the walls of the tunnel with their feet, one foot after the other. The new tunnel was wider, with a towpath, which was built out over the water. The channel therefore extended to the full width of the tunnel and water was able to flow past moving boats and offer little resistance to them.

Coal-mining has caused both tunnels to subside, and the original tunnel was closed to navigation in 1918 for this reason and also because of roof

falls. It remains in existence, though partially blocked, and presumably will continue to do so since both it and its neighbour have been scheduled as ancient monuments. It also continues to perform one of its three original functions by supplying water to the canal: as at Worsley, this contains much ochre and discolours the canal for some distance in each direction. By the time the Brindley tunnel was closed, the Telford tunnel had also subsided to such an extent that much of the towpath was under water: to overcome this problem, and speed two-way traffic through the one tunnel remaining in use, an electric tug service was provided from 1914, and much of the towpath has since been removed to allow boats to steer along the middle of the bore where headroom is least restricted.

By the 1950s traffic was declining and motor boats were common. The tug service was withdrawn, but because of limited ventilation in the tunnel a fan house was built at the eastern portal and completed in 1954. This has three large fans to extract foul air and fumes, and doors to close across the tunnel mouth while the fans are running. Since its completion, boats have passed through under their own power.

Although mining ceased about 1900, subsidence continues.

When I passed through Harecastle in the early summer of 1973 the condition of the bore did not appear excessively bad and a claim by people on a following boat that bricks from the lining had fallen on to it was treated with some disbelief by my crew. Within a few weeks, however, roof falls had blocked the canal. A survey showed seven other sections of the tunnel roof liable to collapse. Repairs involved removal of the tunnel lining, where affected, in eighteen-inch lengths, and its reconstruction, in conditions which were cramped, damp and dangerous; they took until early in 1977.

The passage of Harecastle Tunnel I find, frankly, to be tedious; much water falls from the roof and, if steering, one has to be on the alert not only for this but also for sudden variations in headroom which cause one to duck. It is best enlivened by having someone in the cabin playing the accordion. If Harecastle is tedious, however, Dudley Tunnel (grid ref-

7/5 A day boat laden with passengers emerges from Dudley Tunnel into Castle Mill Basin. Ahead of it another is entering Lord Ward's tunnel.

erence of south portal: SO 933893) is remarkable. One cannot make the passage in one's own boat, unless it is very small (headroom is very much restricted) and one is prepared to leg or pole it through (ventilation is so limited that motors would be dangerous). Fortunately the Dudley Canal Trust operates trips by battery-electrically powered narrow boat. Parts of the tunnel are lined with brickwork and very cramped: but one emerges from these into seemingly vast caverns, wholly artificial, where once limestone was mined. Mysterious branch tunnels disappear towards other limestone

mines and coal-mines, which are now all disused and dangerous—once there were in all some three miles of navigable underground waterway here. Travelling north, one emerges at last into daylight only to find oneself in Castle Mill Basin, surrounded by tree-hung cliffs, from which the boat has to dive into Lord Ward's original tunnel to escape.

Dudley Tunnel's main through route was completed in 1792 and was in continuous use until 1951 when traffic ceased. Its approaches then became derelict and the canal itself in due course a remainder waterway:

happily, following an intensive campaign and much physical work, mostly voluntary, organised by the Dudley Canal Trust with financial contribution by the local authority, the tunnel and its approach canals were reopened in 1973.

The portals of Dudley Tunnel are small, plain, overshadowed by surrounding heights and adorned only with graffiti; they give little indication of what lies behind them. The portals of Sapperton Tunnel on the Thames & Severn Canal were more grandiose, as befitted the principal engineering work of a trunk canal. The eastern portal,

near Coates, Gloucestershire (grid reference SO 966006), has particularly ornate masonry which has been restored by the Stroudwater, Thames & Severn Canal Trust as an early stage in proposed restoration of the entire canal. The tunnel itself is blocked by roof falls.

The longest canal tunnel is Standedge (sometimes spelt Stanedge) by which the Huddersfield Canal pierces the Pennines. Its length is 5,698 yards and it remains in existence, and in water, for the canal, although closed to navigation, is maintained as a water supply channel. Internally it is in bad condition, suffering in unlined sections from falls of rock.

Blisworth Tunnel on the Grand Union is at 3,056 yards the longest canal tunnel still in full use. It is wide enough for narrow boats to pass one another. This tunnel was bored through wet and unstable ground and a first attempt to make it had to be abandoned: it was completed, at the second attempt, in 1805. There is more about this in chapter ten. The nature of the ground through which the tunnel passes still causes problems and the tunnel had to be closed to traffic for the whole of the summer of 1977 while parts of the brickwork lining, which had become seriously distorted, were renewed. Its companion tunnel at Braunston, Grand Union Canal, is closed for repairs as I write, during the winter of 1978–9: after falls of bricks, it was found that much of the lining had deteriorated from the combined effects of air, water and age. This tunnel has a kink in it, due to an error in construction, and it is not possible to see from one end to the other.

Saddington Tunnel on the Leicestershire & Northamptonshire Union Canal was carefully modified during construction so that it could take barges of the type then in use on the Thames which, it was thought, would reach it via the Grand Junction. This was work in vain, for the old Grand Union Canal, which linked it with the Grand Junction, was built with narrow locks, which it retains. The tunnels at the western end of the Trent & Mersey Canal, Barnton, Saltersford and Preston Brook, were also built wide (minimum width of the first two is 13 feet 6 inches and of Preston Brook 13 feet 1 inch) with the intention that flats from the Bridgewater Canal might use them. In practice the transhipment point

between narrow boat and flat became Preston Brook and so far as I am aware these tunnels have never been used regularly by wide craft. An example of a tunnel of exceptionally small bore— about 7 feet wide and 12 feet high—is Morwelldown Tunnel on the Tavistock Canal. It is, however, 2,540 yards long: originally intended for tub boats, it, and the rest of this short canal, were laid out with a slight gradient to cause a flow in the direction used by loaded boats, which would also drive machinery at mines and mills. The boats have long since gone, but the latter function is still performed, in a modernised way, for the canal water now feeds a hydro-electric generating station.

The short Waytown Tunnel on the eastern end of the surviving length of the Grand Western Canal is also of restricted dimensions compared with the grand scale of works farther west: I presume it to have been built for tub boats as part of the canal's short-lived eastward extension to Taunton. A chain ran along the inside to enable boats to be propelled through it: remains of this were present until recently and may be still.

None of these tunnels (except the second Harecastle Tunnel) had or have towpaths. While boats were being propelled through by human muscle, or in later years towed through by tug, horses were led over the top. The horse path is a familiar accompaniment to a canal tunnel: it diverges from the canal itself at the approach to the tunnel and slants away up the side of the cutting to head over the hill on its own. There are good examples at Leek Tunnel on the Caldon branch and Blisworth on the Grand Union. In some places the horse path has become partly or wholly incorporated into the road system of the locality, as at Blisworth, Saltersford and Harecastle Tunnels: above Harecastle it forms a street named *Boathorse Road*.

Shrewley Tunnel on the Warwick & Birmingham (Grand Union) has a curiosity. Approaching the western portal, the towpath diverges gradually up the cutting side: then it enters its own small tunnel, just large enough for a horse. The portal of this is alongside that of the canal tunnel proper, but its floor is level with the crown of the latter's arch. The towpath tunnel continues at an easy upward gradient for a few yards, then steepens sharply: the path soon emerges into a narrow rock cutting and then gains ground level. I deduce that it was laid out like this so that horses might haul their boats as close as possible to the entrance to the canal tunnel; the change in gradient of the horse tunnel probably marks the point at which tow-ropes were detached. The rocky nature of the ground doubtless made it easier to dig a tunnel for the horse path than to excavate a cutting. Further on the horse path becomes typical: a narrow grassy track between hedges, with no obvious involvement with canals.

To continue the towpath through a canal tunnel seems an obvious step, but in the early years it was taken only where tunnels were short—such as Cookley (65 yards) and Dunsley (23 yards) on the Staffs & Worcs and at Armitage (130 yards) on the Trent & Mersey. Cookley Tunnel (grid reference SO 843804) is the oldest surviving canal tunnel which is still in use. Long tunnels were not built with towpaths for some years, until Jessop and Telford provided one through Chirk Tunnel (459 yards) which came into use in 1802.

A much larger tunnel with a towpath was Strood Tunnel on the Thames & Medway, which was exceeded in length only by Standedge, in width only by Netherton, and in height by none at all. I have already mentioned how, later on, its towpath was used for a single track railway and then the whole tunnel for a double track one. Therefore it now appears at first sight a wholly railway tunnel, but its canal origins can clearly be seen by close observation of its approaches. At the Higham or western end, the approaching railway was laid out along the south bank of the canal (which in part survives) and to this day traverses a distinct though shallow S-bend through Higham station to gain the course of the canal before entering the tunnel. At the eastern end, Frindsbury Basin and entrance lock from the Medway are still in existence though disused: standing on the lower gates of the lock (grid reference TQ 743695), which are traversed by a footpath, it can be clearly seen that lock and basin are in a

straight line with the tunnel; but emerging trains bear sharply to the right. As they emerge, the great height of the bore, compared with ordinary railway tunnels, is obvious.

Falkirk Tunnel, on the Union Canal, is the only canal tunnel in Scotland of any length (696 yards); it too belonged to the towpath era. It is still in water and navigable by small craft: much of the interior consists of unlined rock. There were no canal

tunnels in Ireland at all.

Three tunnels only, built late in the canal era, were equipped with dual towpaths. These were Newbold, Coseley and Netherton. Newbold was built as part of the northern Oxford Canal improvements of the early 1830s; since the canal otherwise has but one towpath, each portal incorporates a turnover bridge to give access to the offside towpath through the tunnel—a unique feature. The

7/6 (right) Inside Coseley Tunnel, Birmingham Canal Navigations. Only three tunnels, all built late in the canal era, have dual towpaths.

other two tunnels are both on canals with dual towpaths. Coseley is on the BCN improved main line; Netherton was the last tunnel built in the canal era, opened by the BCN as late as 1858. It is 3,027 yards long and was lit by gas: a marked and spacious contrast with the cramped and overcrowded Dudley Tunnel which it duplicated.

The early canal builders' preference for tunneling rather than making cuttings was so marked that some tunnels passed not far beneath the surface and have subsequently been opened out. The best known example is at Fenny Compton, on the southern part of the Oxford Canal. A mid-way passing-place, open to the surface, was excavated in 1838: then, between 1865 and 1870, the whole tunnel was opened out. That part of the canal is still refered to as *The Tunnel*, and to cruise along it is still remarkably similar to going through a tunnel: the canal, which has wound along the surface of the ground, suddenly becomes narrow and straight as it enters a long gloomy cutting. Rosehill Tunnel on the Peak Forest Canal, immediately north of Marple Aqueduct, was also opened out and is now represented by a narrow cutting, though to judge from appearances, the northern portal remains in existence forming part of a bridge with an unusually high parapet. The short tunnel at Armitage on the Trent & Mersey Canal was opened out as recently as 1971 to counteract mining subsidence.

Aqueducts

It is said that, when water was first admitted to the original Barton Aqueduct, one of the arches appeared to be about to buckle: and that Brindley then retired to bed to think, leaving John Gilbert to have the clay puddle removed and the trough re-puddled with a lesser weight of clay. After which it was sound.

The story may or may not have been embroidered in the telling, but two other points are relevant. The first is that photographs of this stone-built aqueduct taken shortly before it was demolished clearly show arches of brick. Since these are absent from engravings contemporary with the aqueduct when new, which in other respects appear accurate, it seems possible that they had to be inserted later, beneath the original stone arches, to strengthen them. The other is that Barton Aqueduct had the appearance of a very much more slender structure—so far as any stone aqueduct can be slender—than the aqueducts which Brindley built next.

Barton Aqueduct looked slender partly because its piers had to be high to provide headroom for flats on the Mersey & Irwell Navigation below; Brindley's next aqueducts were over un-navigable rivers, such as the Dove, which the Trent & Mersey crosses north east of Burton, and the Trent, which the Staffs & Worcs crosses at Great Haywood. These aqueducts are low and massive, the arches small, the superstructures substantial: almost a row of culverts.

Later engineers gained both knowledge and confidence. Robert Whitworth, Brindley's pupil, provided the

Forth & Clyde Canal with its greatest engineering work in the form of Kelvin Aqueduct near Glasgow. It has four arches, a total length of 400 feet, and a maximum height of 70 feet; it was completed in 1790. John Rennie, the most architecturally-conscious of canal engineers, was to provide canals with their most stylish of aqueducts, notably those which carry the Lancaster Canal across the River Lune near Lancaster, and the Kennet & Avon's Dundas and Avoncliff Aqueducts across the Avon. All display the features of contemporary classical architecture, and have massive cornices; the Lune Aqueduct is topped by elegant balustrades. Dundas and Avoncliff Aqueducts were built of Bath stone.

The Lune Aqueduct was completed first, in 1797; it has five semi-circular arches, is 60 feet high and 600 feet long. Avoncliff (grid reference ST 805600) and Dundas (grid reference ST 785625) Aqueducts were completed in 1798 and 1800 respectively; each has a principal arch across the river, flanked by lesser ones. At later dates the approach embankments of both these aqueducts were pierced to allow a railway to pass beneath the canal.

In Ireland the Grand Canal crosses the River Liffey near Naas by Leinster Aqueduct, a substantial five-arched stone structure built in the early 1780s. It had originally been intended to cross the river on the level, by locking down into it and up again on the far side.

This method was in fact adopted at Monasterevin, where the Barrow Line of the canal crosses the River Barrow. The present aqueduct was not completed until 1826, and with an air of lightness provided by comparatively flat arches and white-painted iron railings along the towpath, its appearance confirms it to have been a more recent structure than Leinster Aqueduct. Traces of the original route down into the river, including a blocked-up road bridge, could be seen until recently and probably still can.

The Royal Canal had more broken country to traverse than the Grand Canal, and its engineering works tend in consequence to be on a larger scale. The three-arched Boyne Aqueduct,

which carries this canal across the River Boyne near Longwood, Co. Meath, is higher than any on the Grand Canal, and overshadows, figuratively if not literally, the adjacent railway viaduct. The approach embankments, too, are both long and high. The canal here is still in water, though weed-grown.

As canals penetrated more and more difficult country, so it became desirable for aqueducts to be loftier, to span not only river but ravine as well. This meant that reducing the weight of the superstructure became important, otherwise very thick piers were needed which, as aqueducts became taller, became expensive in proportion. Benjamin Outram, engineer of the Peak

Forest Canal, made a contribution towards solving this problem by building Marple Aqueduct (grid reference SJ 955901) with the spandrel walls pierced by cylindrical cut-outs. This three-arched masonry aqueduct carries the canal almost 100 feet above the River Goyt, and was built between 1795 and 1800.

In 1962, during the period when the canal had become un-navigable, part of the side of the aqueduct collapsed, and water was piped across. Scheduling of the aqueduct as an ancient monument in 1963 at the instance of the local authority was followed by a grant from Cheshire County Council towards the cost of repairs. This action was instrumental in keeping the canal

7/7 Boyne Aqueduct by which the Royal Canal crosses the River Boyne near Longwood, Co. Meath, is little known, although one of the largest in Ireland.

Tardebigge top lock, Worcester & Birmingham Canal, has at fourteen feet probably the greatest fall of any narrow lock. It occupies the site of an experimental vertical lift built in 1808—by use of such lifts, the canal company hoped to reduce greatly the number of locks needed for the descent to the River Severn. But although the lift operated satisfactorily on trial, it was not considered sufficiently robust for general canal use, and locks were eventually built—fifty eight of them in sixteen miles, making one of the most heavily-locked sections of canal in the British Isles.

The Hay inclined plane, Coalport, on the Shropshire Canal. Tub-boat canals solved the problem of providing canal transport in hilly districts—tub boats, very small, were towed along the pounds several at a time and raised or lowered singly from one pound to another by inclined planes. On these they were carried on cradles which ran on rails. The Shropshire Canal was opened in 1792, and The Hay inclined plane was last used about 1894. Its site then lay derelict for many years, but railway track, and water in the lower pound of the canal (foreground) have recently been reinstated by the Ironbridge Gorge Museum Trust. The track is of the usual railway type, for the canal became part of the Shropshire Union Railways & Canal Company's system and its inclined planes were re-laid, in their later years, with ordinary railway track.

intact for eventual reopening.

The problem of carrying a canal across deep valleys became acute during construction of the Ellesmere Canal, the main line of which, intended to run from the Mersey to the Severn, had to cross the valleys of the Dee near Ruabon and the Ceiriog near Chirk. It was solved by the use of cast iron in the spans: a solution which was to provide canals with their best-known and most impressive structure—Pontcysyllte Aqueduct (grid reference SJ 271420). This has nineteen spans, is 1,007 feet long, and carries the canal at one point 127 feet above the River Dee.

The original intention of the promoters was to cross the Dee by a low level masonry aqueduct, to each end of which the canal would descend by flights of locks. That was early in 1793, before Telford was appointed engineer under William Jessop. Use of cast iron was eventually to enable the canal to be built at high level without locks, though just how the idea of using it at Pontcysyllte arose is not at this distance of time clear.

What is known is that in February 1795 Telford was also appointed engineer of the part-built Shrewsbury Canal, following the death of its original engineer Josiah Clowes. This canal had to cross the little River Tern at Longdon-upon-Tern (grid reference SJ 617157) but a masonry aqueduct which Clowes had been building had been swept away by floods. The ironmasters of Shropshire were prominent in the Shrewsbury Canal company and it was from them that the idea came of building an iron aqueduct at Longdon. A fortnight after Telford's appointment it had been approved.

The ends of Clowe's aqueduct were re-used, to become massive abutments pierced by small brick arches: between them in total contrast the iron aqueduct was built, light and airy by comparison. Not only the trough but also its supports were made of iron; the towpath was positioned level with the base of the trough and the overall width of trough plus towpath was less than one third of the original. The aqueduct is 62 yards long and 16 feet high; it was completed in 1796 and remains in existence, an ancient monument, though since the canal is abandoned it is now waterless and even the approach embankments have been demolished.

It may be that the decision to use iron at Longdon gave Telford, whose training had been as a stonemason, the idea of using it at Pontcysyllte. Or it may be that, as an intelligent man and, since 1787, Surveyor of Public Works in Shropshire, where the early ironworks of the industrial revolution were situated, he was aware of the potential of cast iron and had been considering it for some time, finding at Longdon an

7/8 Cast iron meets masonry at Longdon Aqueduct, Shrewsbury Canal (grid reference SJ 617157). Parts of Clowe's original aqueduct still stand, but the central portion, washed away by flood, was replaced by Telford with cast iron spans and supports – the first use of cast iron for an aqueduct.

opportunity to demonstrate his ideas in practice. Or yet again, the idea for iron at Pontcysyllte may have originated with William Jessop: Jessop was a partner in Butterley Ironworks with Benjamin Outram, who between 1793 and 1796 built a small iron aqueduct on the Derby Canal. Yet Outram rejected iron for Marple Aqueduct, although it was considered.

Probably the truth lies in the midst of these extremes, in an interplay of men and ideas. In any event, it was Jessop, the Principal Engineer, who formally recommended to the Ellesmere Canal committee in mid-1795 that a high-level aqueduct with iron spans should be built at Pontcysyllte, and Telford the subordinate who prepared detailed plans and, in the initial stages, supervised construction.

Before much work had been done, however, and that only on the piers, the committee decided to concentrate on the aqueduct across the Ceiriog at Chirk (grid reference SJ 286372) with a view to opening that section of the canal first since it promised revenue from local traffic. Chirk Aqueduct was built between 1796 and 1801 and spans wholly of iron were not used. It may well be relevant that work started before Longdon Aqueduct was complete, or very shortly afterwards, and the principle of large iron trough aqueducts had not then been demonstrated to be practicable. What was done to lighten the structure was this: the bed of the trough was made of inch-thick cast iron plates, and was supported from each arch not by solid masonry but by five parallel lengthwise walls with spaces between them. The sides of the trough were made of masonry: the top surface of one of them became the towpath, fenced by an iron railing. This construction for the trough was evidently not watertight, or unsatisfactory in some other way, for the iron bed was subsequently removed and a complete iron trough inserted: this is still in use.

The aqueduct does, however, have a problem: over the years, rainwater has percolated the towpath, and the masonry of the opposite side of the trough, washing away mortar from internal and external walls and arches and causing the arches to settle. The method to be adopted for repairs has not been announced at the time of writing.

Work on Pontcysyllte Aqueduct re-

7/9 Chirk Aqueduct, Ellesmere Canal, seen from the west side. Immediately beyond is the entrance to Chirk Tunnel.

started in 1800 as Chirk neared completion. By now, however, the euphoria of the Canal Mania had long since evaporated, and the company decided not to build its line onwards from Ruabon to Chester: a tramroad would connect industries in the Ruabon area with the canal, and the company even considered terminating the canal south of the aqueduct and using this structure for the tramroad instead. Fortunately for later canal enthusiasts it reconsidered, and the aqueduct was completed as such in 1805. The main line of the canal terminated 300 yards north of it, the line to Llangollen being a branch.

Telford had sole responsibility for construction of the greater part of Pontcysyllte Aqueduct, for Jessop had

ceased to be engineer to the company about 1801. As eventually built the aqueduct has tapering piers of stone, hollow towards the top for lightness, and spans of cast iron. The internal width of the trough is 11 feet 10 inches and the towpath is cantilevered above the water, leaving sufficient width for narrow boats. Here Telford adopted the same technique, which allows water to flow freely past moving boats, that he was later to employ in his new Harecastle tunnel.

Each span of the trough is supported by four parallel arched cast iron ribs. A routine inspection in June 1975 revealed that in the southernmost span, adjacent to the abutment, ribs were badly distorted and some were fractured, and there were large cracks

Several horse-drawn passenger trip boats operate on various parts of the canal system, and provide opportunities to see boat horses at work. Here, beside the Shropshire Union Canal at Chester, the mare Snowy *is being prepared to haul the passenger narrow boat* The Chester Packet *in the background. Brightly coloured bobbins on her harness are traditional, and prevent the traces from chafing her sides.*

Authentic traditional paintwork by Ron Hough is displayed on the narrow boat Raymond. Raymond *was the last wooden narrow boat built, and was launched in 1958 at Braunston: both her design, and the methods used to build her, represented a continuous tradition by then almost 200 years old. Unpowered, she was used as butty to a motor boat by Samuel Barlow Coal Co. to carry coal between the Warwickshire coalfield and London. As part of Barlow's fleet she was later taken over by Blue Line and continued to operate until the traffic ceased in 1970. She is now moored permanently at Braunston, where her captain Arthur Bray still lives aboard but no longer travels.*

7/10 (p.111-112) Primaeval canal bridge, narrow with no towpath, at the tail of Woodend Lock on the Trent & Mersey Canal, was built c.1769. See also illustrations 7/11 and 7/12

Old liveries have been restored to exhibits at the Boat Museum, Ellesmere Port. Horse-drawn narrow boat Gifford—*her hold is decked-in to carry tar or oil—has been restored to the colours of her original owner, Thomas Clayton (Oldbury) Ltd, who had a fleet of such boats, and in the background is Fellows, Morton & Clayton Ltd narrow boat* Monarch, *which was originally steam-driven and has been restored externally to her original colours and appearance.* Gifford *was built in 1926,* Monarch *in 1908.*

Transom-sterned wooden Leeds & Liverpool Canal short boat George *is a rare survivor of a type of boat once common on that canal. The keelson made from a metal girder is typical of the type.* George *is an exhibit at the Boat Museum and is here seen in a dry dock formed of the short pound between two wide locks at Ellesmere Port. Parallel narrow locks can be seen on the right, and between wide and narrow locks a long low roof supported on posts shelters a set of stop planks. The building with the chimney in the distance is a pump house which generated hydraulic power for the docks' hydraulically operated cranes and capstans.*

between ribs and masonry. The cause was eventually traced to slight rotational movement of the abutment: records at Stoke Bruerne museum showed that there had been similar trouble previously, in 1866. The canal across the aqueduct had to be drained for repairs which involved supporting the trough temporarily by timber trestles, stabilising the abutment and replacing the ribs: simpler to describe than do—the canal was closed for the rest of the 1975 cruising season and boats trapped the wrong side of the aqueduct had to be craned out and moved elsewhere by road. One of the damaged ribs is now an outdoor exhibit at the Waterways Museum, Stoke Bruerne.

Of other iron trough aqueducts, the wide aqueduct on the Grand Junction Canal over the River Great Ouse at Wolverton was built in 1811, replacing an earlier structure which collapsed. The southern section of the Stratford Canal has three iron trough aqueducts, of which the largest, Bearley or Edstone Aqueduct, is over 475 feet long and between 19 and 28 feet high. Although built during the period 1812–6, its appearance is old-fashioned for the date, more reminiscent of Longdon than Pontcysyllte. The towpath runs level with the base of the trough, which is little wider than a narrow boat.

An aqueduct of unusual and unique design was built at Stanley Ferry and completed in 1839 to carry the Aire & Calder's improved line to Wakefield over the River Calder. It crosses the river with a single span of 164 feet: the cast iron trough, 24 feet wide, is supported on cast-iron cross beams which are in turn hung by wrought iron hangers from two cast iron arches which span the river. The arches spring from masonry abutments below the river's usual water level. In 1974 consulting engineers pronounced it to be beyond economic repair for operational use, and a new aqueduct, to stand alongside it, was designed. The original, an ancient monument, is to be preserved.

The most impressive and interesting group of aqueducts after Pontcysyllte and Chirk is undoubtedly the three large aqueducts of the Union Canal: Avon, Slateford and Almond. They are respectively 810, 500 and 420 feet long and 86, 75 and 76 feet high; Pontcysyllte and Chirk for com-

parison are 1,007 and 600 feet long and 121 and 70 feet high. The Union Canal aqueducts were designed by Hugh Baird, the canal's engineer, with advice from Telford, and the design is that not of Pontcysyllte but of Chirk Aqueduct, as modified: that is to say they are masonry aqueducts (and very fine masonry too) with iron troughs for the water channel. They are, however, wider than Chirk, being intended for barges 12 feet 6 inches wide rather than narrow boats; and their arches have a much greater span. They were completed in 1822. They seem to have antedated the modification to Chirk Aqueduct for according to E. A. Shearing, writing in *Waterways News* (April 1979), Chirk Aqueduct was not rebuilt until 1869.

All the aqueducts so far described carry canals over rivers or valleys or both. Perhaps more romantic are those aqueducts which carry one canal above another, and evoke those distant times when small canals, far from being a forgotten means of transport, flourished so increasingly that there was trade for two canals at different levels in the same district.

They are to be found, mostly, on the Birmingham canals. Stewart Aqueduct (grid reference SP 003898, aqueduct obscured by M5 motorway) carries the old main line over the new, and Tividale Aqueduct (grid reference SO 968909) carries it over the approach to Netherton Tunnel. Engine Arm Aqueduct, all Gothic ironwork, carries an important feeder canal across the new main line to the old.

There are two curious locations on the Trent & Mersey where the lie of the land makes it necessary for a branch, which is to head in one direction from the main canal, to diverge actually from the opposite side of it, run parallel while the main line falls through locks and then, where the valley narrows, cross over it to gain its intended direction. These are at Red Bull Aqueduct where the mile-and-a-half long Hall Green branch (which leads to the Macclesfield Canal but was built, for reasons of canal politics, by the T & M) crosses over the main line, and Hazelhurst, where the Leek sub-branch crosses over the Caldon branch. Barton Swing Aqueduct carries the Bridgewater Canal over the Manchester Ship Canal: it is referred to in detail in chapters six and ten.

Aqueducts built to enable railways

to pass beneath canals were unusual. The Grand Western Canal has one, built with two arches for double track, although so far as I am aware only one track was ever laid and even that has now been lifted since the line concerned, the Tiverton branch, has been closed. This is one of those ironic instances of places where the canal has survived and outlived the railway which once seemed all-conquering; Brecon is another. I digress, though there are other instances of canal-over-railway aqueducts where the railway is much less busy than formerly. Two separate railways were laid out so as to cross the Stratford canal beneath Edstone Aqueduct, though only one of them survives, and the Cromford Canal near High Peak Junction crosses over the Matlock branch which was once a main line to Manchester.

The points of diminishing size at which an aqueduct over a stream becomes a culvert and an aqueduct over a road becomes an underbridge are imprecise. Culverts are inevitably commonplace on all canals and one is tempted to define them as water channels passing through the foot of a canal embankment, and aqueducts as being structures (rather than earthworks) which themselves carry a canal. The distinction falls down, however, on late canals such as the Macclesfield where the builders, having perfected cut-and-fill technique, formed some very large bridges in their embankments, to take rivers, streams and roads beneath the canal. These structures, in places more than half the height of the embankment, seem more appropriately to be described as aqueducts than culverts or bridges. There is a good example on the Macclesfield Canal at Sutton Aqueduct, close beside James Brindley's place of apprenticeship (see chapter three): the canal embankment is pierced by two large arches at different levels, the lower spanning the River Bollin and the upper, close to it, still of ample size, even now, for a busy road.

As mentioned earlier, there are many places on the Runcorn line of the Bridgewater Canal where roads pass beneath it by 'underbridges', but this technique was uncommon throughout most of the canal era. Among the few exceptions are an underbridge on the Peak Forest Canal near Romiley and a few on the Grand Junction and the western end of the Trent & Mersey.

7/11 (above) From the middle years of canal construction a typical humped bridge on the wide Grand Junction Canal north of Linslade, built c.1798.

7/12 (below) Fine masonry and design on a bridge of the Macclesfield Canal built c.1828. But the canal is narrow, for the time had passed when a network of wide canals seemed likely to be built.

Blundell Aqueduct on the Grand Canal enables a main road to pass beneath the Bog of Allen embankment near Edenderry, and is so long, from the road user's point of view, that it is called *The Tunnel*. The southern Stratford Canal is carried over a main road at Wootton Wawen by an iron trough aqueduct, and the Ashton Canal's Store Street Aqueduct, a masonry structure, carries the canal over the Manchester street of that name.

It was only towards the end of the canal era, when canals were being built along direct lines by cut-and-fill, that it became as common to carry canals above roads as beneath them. On the Union Canal underbridges, or small aqueducts over roads etc., come as frequently as underbridges on a railway; much the same is true of the Macclesfield and the Birmingham & Liverpool Junction Canals, and there are several such works on the Oxford Canal cut-offs. On the latter two canals are several iron trough aqueducts of more refined appearance than Wootton Wawen; that at Stretton, where the B & LJ crosses over the main road now called the A5, is particularly fine. It is good to see that in recent years hoardings which disfigured it have been removed, and nondescript grey paint has given way to jet black with features picked out in white, especially the cast-in lettering *Birmingham and Liverpool Canal, Thomas Telford Engineer*.

Bridges

The most characteristic of all canal features is the humped bridge. Some are of brick, some of stone, some are rendered, some are not. Some are part brick, part stone. Early brick bridges were built of red bricks baked from clay found along the course of the canal, later ones have their features picked out in blue engineering bricks. Stone was used wherever it was available—Cotswold stone on the southern Oxford Canal, grey limestone in Ireland, millstone grit on the Leeds & Liverpool. All humped bridges have that characteristic shape: a narrow arch over the canal surmounted by the wider arc traced by the rim of the parapet—although, as I shall mention shortly, in bridge design, as in other canal features, there was much development during the canal era.

Canal bridges were built for three purposes: to enable the public to cross the canal, by road, bridleway or foot-path; to enable riparian landowners, whose land was severed by construction of the canal, to cross over it on 'accommodation bridges'; and to enable those people, and their horses, whose business was with the canal itself to cross over it. This meant bridges to carry the towpath from one side of a canal to the other and, at locks, foot-bridges for people working them. Some bridges combined two or three of these purposes.

The characteristic shape arose from the need to provide headroom for boats under bridges which crossed canals laid out at ground level. Tub boat canals, for small boats without cabins, had noticeably limited headroom under bridges. Generally, through a bridge hole, the channel was made little more than wide enough for a single boat, though some narrow canals were built with wide bridges, more for ease of navigation and the possibility of future widening of the canal than to enable boats to pass one another beneath them; in any event, the normal width of the canal is usually restricted through bridges. Both of these factors meant that it was convenient to site bridges, where possible, at the tails of locks, or alternatively to position locks immediately on the uphill side of public road bridges, and so to make the masonry or brickwork continuous throughout the combined structure.

On early canals, detail design of bridges and other small structures was very much at the whim of the actual builder of a particular section of canal; later on it became to a greater extent the responsibility of the canal company's engineer, and more uniform structures were provided. The canal bridge at its most primaeval is to be seen at the tails of some of the locks by which the Trent & Mersey Canal climbs through Alrewas and Fradley. The bridge hole is a low, narrow opening in the brickwork, the same width as the lock chamber: to pass through it, boats had to have their tow-ropes detached. However, so that horses of approaching boats could walk up alongside the lock, and tow-ropes be detached at the last moment, the parapets on the towpath side curve round and descend gradually and completely to ground level so that tow-

ropes may rise up on them without becoming snagged. Close inspection of some of these bridges leads me to suspect that even this feature was lacking at first and may be a modification of the original design.

There are other instances of bridges across the tails of locks, with the consequent inconvenience of detaching tow-ropes. Sometimes they are footbridges for people working locks, and sometimes they carry public roads. They are to be found on the Staffs & Worcs and the Leeds & Liverpool, for instance, and on some other canals built surprisingly late in the canal era, such as the Peak Forest and the original Grand Union. There are at least two instances where a little horse tunnel was built beneath a road, adjacent to such a bridge: these are at Stone on the Trent & Mersey and Marple on the Peak Forest. In each case I deduce that the road concerned was a busy one, even at the date when the canal was built, and that the horse tunnel was provided so that horses trailing ropes did not have to tangle with the traffic; but why the seemingly obvious solution of carrying the towpath beneath the canal bridge was not adopted I do not know.

At any rate it was not long before the towpath was generally taken through bridge holes wide and narrow, whether at the tails of locks or not. Early wide-canal bridges can be seen on the Bridgewater Canal, and the Chester Canal section of the Shropshire Union. The outlines of their parapets and bridge holes are often irregular: later bridges have smoother curves, and better proportions which combine utility with largely unconscious elegance.

At first canal bridges were built at right angles to the canal: so in many places roads had to make inconvenient Z-bends across bridges. So that roads not at right angles to canals might cross without diversion, skew bridges were introduced. The arch of a skew bridge has winding courses of brickwork or masonry, as in plate 7/13, to avoid the lines of weakness which otherwise occur across the bridge's width, and at the same time to enable the arch to be made to the minimum span needed for canal and towpath. The first skew bridges were probably built in Ireland on the short Kildare Canal which was opened between Naas and a junction with the Grand

7/13 A masterpiece of the stonemason's art: winding courses of a skew bridge on the Macclesfield Canal.

Canal near Sallins in 1789. The engineer Richard Evans had been requested by his directors to preserve the angle of roads crossing the canal. In 1808 the canal was purchased by the Grand Canal company, and as these bridges lacked towpaths and were lower than Grand Canal bridges, they were rebuilt as ordinary bridges. I understand traces of the original skew construction can still be seen.

The first skew bridge in England was probably March Barn Bridge, built across the Rochdale Canal in 1797 near Castleton, Rochdale (grid reference SJ 886111). Later they became commonplace on, for instance, the Grand Junction Canal and the Birmingham & Liverpool Junction. Skew bridges at their most extreme are to be seen on the Macclesfield Canal: they are superb examples of the stonemason's art. Nevertheless, so far

into obscurity did canals later fall, that it was commonly thought that skew bridges originated with railway engineers.

Where the Birmingham & Liverpool Junction section of the Shropshire Union passes through its deep and narrow cuttings, such as Woodseaves and Grub Street, it is crossed by 'high bridges': parallel-sided masonry abutments tower above the canal and are spanned by arches of traditional type.

Some bridges over Telford's new main line of the Birmingham Canal may be taken as the traditional canal bridge at its most highly developed. Here, the canal does not narrow through bridge holes: the lines of the towpath edges on either side are not interrupted. Where the canal is at ground level, this was achieved by building double-arched bridges with a central island pier; the parapet de-

scribes a single shallow arc above them. Where the canal is in a cutting, large brick skew arches span the entire width of the canal and towpaths. Examples are described in chapter ten.

Deliberate ornamentation of canal bridges was rare, but none the less pleasant where it occurs. Parapets of bridges on the Pocklington Canal terminate on either side with neat brickwork columns. Some stone humped bridges on the Union Canal have iron railings in place of the usual solid parapets. Some bridges were ornamented to embellish a landowner's park, such as Avenue Bridge on the B & LJ near Wheaton Aston, which has rusticated masonry topped by balustrades, and Solomon's Bridge at Cosgrove on the Grand Junction which has Gothic features. Other ornamental bridges were built thus to enhance city amenities. Such are Hubard Bridge over the

118

FEATURES OF CANALS: THE WATERWAY

Grand Canal in Dublin and Macclesfield Bridge across the Regent's Canal in London. The latter is an unusual design: the canal is in a cutting on the north side of Regent's Park and the bridge's shallow brick arches are supported on rows of cast iron columns either side of the canal. The bridge is famous for its destruction in 1874 by explosion of a barge carrying petroleum and gunpowder: the iron columns survived and the bridge was rebuilt.

Several bridges have already been referred to by name but the general practice in Britain is for canal bridges to be identified by number, the names, if any, being supplementary. Generally the numbers were and are carried on cast iron bridge plates positioned above the arches; occasionally they were carved in stone. The Staffordshire & Worcestershire and the Birmingham & Fazeley Canals were exceptions: their bridges were all named, and carry their names on the bridge plates. Generally the names relate to the vicinity or to personalities now forgotten (who, one wonders, was Long Moll, who had a bridge named after her on the Staffs & Worcs?). On the Grand and Royal Canals in Ireland, too, bridge names rather than numbers are the rule: several of the names commemorate canal company directors (such as Binn's Bridge, and Huband Bridge just mentioned). Others again are geographical. Bridges bear their names carved in eighteenth-century script on large stones built into the parapet above the crown of the arch.

Canal bridges were built to carry the traffic of the time—people, animals, light horse-drawn vehicles. The brick arch bridges of the Stratford Canal, for instance, have a loading of five hundredweight. This meant that later on many bridges came to bear caution notices advising drivers of traction engines and other ponderous carriages that the loads they could bear were restricted. Diamond-shaped cast iron notices on posts, facing the traffic either side of the bridge, carried a similar message. Notices of both types survive in service, bearing the names of long-superseded canal companies or (to the bafflement, presumably, of the ordinary passer-by) canal-owning railway companies. Since 1968 many canal bridges which carry public roads have been strengthened or rebuilt at

Government expense to carry present-day traffic—part of a wider exercise called Operation Bridgeguard. The cost of maintaining these bridges in their improved condition will, however, eventually revert to the navigation authority.

Despite that, some canal bridges have proved strong enough to bear main road traffic far heavier and more intensive than their builders could have imagined, and have been widened once if not twice in more modern manner. Examples of such bridges extended (from the canal point-of-view) may be seen at Curdworth on the Birmingham & Fazeley and Nell Bridge near Aynho on the Oxford Canal. The latter has become particularly awkward for the boater for the original bridge, if I remember rightly, is at the tail of a lock and has no towpath so that, although later extensions do not affect the width of the channel, their effect is that the towpath terminates abruptly in the dark beneath the bridge, and people put ashore to work the lock have to emerge from it and brave the traffic on the A41.

Roving or turnover bridges carry the towpath across a canal. The terms are sometimes treated as interchangeable, but properly speaking a roving bridge carries the towpath of one line of canal over another at a junction, while a turnover bridge carries the towpath from one side to the other of its own canal. The latter are often located near aqueducts for, where a contour canal crosses a valley, the towpath needs to be transferred from one side to the other so as to remain on the downhill side of the canal.

The distinctive feature of almost all turnover bridges is that on one side of the canal the towpath continues beneath the arch and only then ceases: horses first passed beneath the bridge, then turned up a ramp to cross over the bridge and gain the continuing towpath the other side. Like this, there was no need to detach tow-line from boat. Parapets, of course, all originate at ground level and rise gently to offer a snag-free run for tow-lines. The incline up which a horse can walk being only a gentle one, some turnover bridges have two matching straight ramps along the canal: the whole a pleasantly symmetrical structure. There is a good example on the Oxford Canal at Braunston, the first bridge south of Braunston Turn.

Elsewhere, the towpath under the bridge leads into a ramp which spirals up on to the bridge, accompanied on each side by a spiralling wall to keep horses from straying from the correct route. Bridge no. 105 on the Staffs & Worcs at Milford, near Stafford, is an early example. The stone turnover bridges of this type on the Macclesfield Canal are particularly well-proportioned and pleasing to the eye (grid references of typical examples: SJ 962883 and SJ 925719). The Macclesfield Canal, though built by cut-and-fill, follows the western slopes of the Pennines and throughout most of its length the usual practice of placing the towpath on the downhill side was followed. Through towns, however—Congleton, Macclesfield, Marple—the downhill side is occupied by wharves and the towpath is transferred to the opposite bank (so that moored boats would not obstruct tow-lines, so that goods on the wharves would not impede horses, so that goods would be secure from passers-by). At each of the approaches to these towns therefore there is a turnover bridge: and to simple reasons of operating convenience we owe what are perhaps the most attractive small structures on the canals.

There is what might be called a Mark I version of these bridges at Hyde on the Peak Forest Canal, but, though built of stone and of similar layout, its lines are crude by comparison. At Chester, over the Shropshire Union close by its junction with the Dee branch, is an attractive and unusual turnover bridge which combines an arched span of iron, surmounted by iron railings, with a brick-built spiral.

At Fenny Compton, the towpath of the Oxford Canal formerly crossed from one side to the other by means of the horse path over the tunnel. Opening out the tunnel meant that a turnover bridge had to be provided in the resulting cutting and this remains in situ. A pleasant cast iron structure, it has the curious feature that the towpath does not pass beneath it and tow-ropes must therefore have been detached for boats to pass it. One hopes that boatmen doing so did at least consider this a lesser chore than legging through a long tunnel!

One of the earliest roving bridges is that which carries the Trent & Mersey towpath over the Staffs & Worcs Canal

119

and its towpath at Great Haywood Junction. Built of brick, it is slimmer and flatter than most humped bridges. There is a tendency for bridge holes of roving bridges to be wider than most, doubtless to enable boats to manoeuvre round the corner. Although many roving bridges were built of brick or stone—those at Fazeley and Autherley Junctions are good examples—it is those made of cast iron which are notable. The Birmingham Canal Navigations and the Oxford Canal made great use of them, particularly to carry the towpaths of their improved lines over the spurs, loops or branches which were the remains of the old. They were cast in sections and bolted together: they spring from brick abutments and have a graceful shallow arch, and railings too were made of cast iron, in ornamental patterns.

Use of cast iron enabled another type of canal bridge to develop: the 'split' or divided bridge. They were built only on narrow canals and on close examination prove to comprise

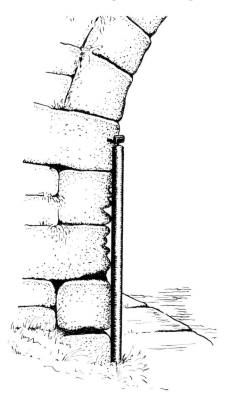

7/14 The grooves worn in the corners of the arches of canal bridges by wet, grit-impregnated tow-ropes are a familiar sight. Often the corners were protected, as in this instance on the Rochdale Canal, by a vertical roller.

two abutments a little over seven feet apart upon which are mounted iron brackets for the deck: these fail to meet across the canal by a gap sufficiently wide for a tow-rope to pass through, but no wider. The width of the bridge, and therefore its cost, were minimised, as the towpath does not have to pass beneath it. This type of bridge is most common on the southern Stratford Canal where it was adopted for many of the accommodation bridges, but it was used by many other canals also, particularly the Stourbridge and the Trent & Mersey for footbridges at the tails of locks. I cannot call to mind any instance where it was used for a public road.

At the other end of the scale, however, one of the greatest bridges on the canal system was built of cast iron. This is Galton Bridge, built by Telford in 1829 to carry Roebuck Lane over the new BCN main line near Smethwick. The canal here is in Galton Cutting, and a tracery of ironwork carries the road across it at high level, some seventy feet above the water.

A well-known characteristic of canal bridges is the grooves worn in the corners of the arch by the abrasion of countless tow-ropes, wet and impregnated with towpath grit. It is intriguing to watch the tow-line of a present-day horse-drawn tripping boat and see that, sure enough, when the horse passes beneath the bridge, the moving rope takes up its position in one of these grooves. In many places the corners of bridge arches were protected by cast iron plates, which themselves became grooved in time, or by rollers. The railings of roving bridges, and divided bridges too, are in some cases badly eroded by tow-lines.

Since the canal era some canal bridges have had, from time to time, to be replaced in the interests of increasing road traffic. Such a one was Broad Street Bridge over the BCN at the top of Wolverhampton locks, which was rebuilt in 1878 and has now in turn been replaced by a modern concrete bridge. The 1878 bridge has been removed and re-erected in the Black Country Museum. The end plates of the bridge, as seen from the canal, are of cast iron, but the main structure is of wrought iron. Recent road bridges are generally of concrete: some of these were built early enough to have to cater for horse-drawn boats and therefore include the anachronistic features of paved and

cross-ribbed towpath, and rubbing plates let into the concrete of the abutments to prevent tow-line abrasion. An example I noted in 1978 was Henhull Bridge, no. 95, on the Shropshire Union near Nantwich.

Timber was little used on canals for fixed bridges other than footbridges: and in these, concrete spans have in many cases replaced the original timber ones. One wooden trestle accommodation bridge that does survive crosses the Trent & Mersey Canal at Wychnor, east of Alrewas.

Where timber came into its own, however, was in construction of opening bridges. These were cheaper than fixed bridges, both intrinsically and because the approach roads did not have to be elevated, and were used on many small canals for accommodation bridges and bridges carrying unimportant roads; they were essential for all bridges on ship canals used by craft with masts. Although many opening bridges have had to be replaced by more modern structures, there are still a great many which appear to be contemporaries of canal construction in design, even if some of their component timbers are replacements.

Opening bridges come in two classes: lifting bridges which open vertically and swing bridges which open horizontally. The former can be subdivided further into drawbridges, bascule-bridges and lift bridges. In drawbridges a deck hinged to one abutment is suspended at the far end from horizontally pivoted overhead balance beams which have to be pulled down to raise the bridge. Such bridges are common on the Ellesmere Canal, and found less frequently elsewhere on, for instance, the Caldon branch of the Trent & Mersey and the Northampton branch of the Grand Union. One of the few opening bridges on the Grand Canal in Ireland is of this type, at Monasterevin on the Barrow Line.

Second, and less common, are bascule-bridges in which a hinged deck is balanced against a counterweight. Those on the southern part of the Oxford Canal are best known: the timber deck is balanced against timber balance beams, of the proportions of balance beams of lock gates, which are pulled down to open the bridge. This type of bridge is, strictly speaking, I suppose, a simple type of rolling lift bridge, for the pivot comprises a

toothed cast iron segment working in a matching rack. The Forth & Clyde Canal was originally equipped with wooden double-leaf bascule-bridges of which a few examples survive, though no longer in use as opening bridges, at Craigmarloch near Kilsyth, for example, and at Bowling. Increasing road traffic meant that some of the original bridges were later replaced by steel swing bridges and at least two remain, but are no longer swung.

Lift bridges are the third type of bridge to open vertically: the entire deck is raised, remaining horizontal, by means of a system of chains or ropes, pulleys and counterweights. They are rare on canals. One example has been re-erected in the Black Country Museum: it originally crossed the entrance to a railway-interchange basin. Another carries Leamington Road across the Union Canal close to its terminus in Edinburgh: it was formerly electrically operated and, though unusable at present, there are plans for it to be restored.

Swing bridges are widespread on canals; early ones were built of timber, later ones of steel. They are common on the Leeds & Liverpool Canal and the Kennet & Avon; they are also to be found on the Ashton, Peak Forest, Macclesfield, St Helens and Grand Union Canals. Drayton Manor swing bridge on the Birmingham & Fazeley is accompanied by an ornamental foot-bridge with a wooden span supported by two small brick battlemented towers. Some swing bridges cross locks. There is an example on the St Helens Canal at Bewsey Lock (grid reference SJ 593896) which remained in existence in 1978, and another which spans what was formerly a stop lock at Great Northern Basin at the head of the Erewash Canal. This bridge was restored in 1973 and made to operate again after some thirty seven years of disuse.

Larger swing bridges were installed on ship canals such as the Gloucester & Berkeley and the Caledonian. The Caledonian was equipped originally

7/15 (above) A typical swing bridge, on the Leeds & Liverpool Canal near Silsden.

7/16 (right) Timber drawbridges were cheaper to build than fixed bridges, and are locally common on the Llangollen Canal. This one is at Whixhall Moss.

7/17 (overleaf) Moy Bridge, near Gairlochy, is the last of the cast-iron double-leaf swing bridges with which the Caledonian Canal was originally equipped. Grid reference NN 163826.

121

with double-leaf swing bridges of cast iron. All but one have been replaced by more modern bridges: the survivor is an accommodation bridge at Moy near Gairlochy (grid reference NN 163826) which was built about 1812. It has been scheduled as an ancient monument, one which happily remains in use.

Where single-leaf opening bridges were built, the pivot or hinge was almost invariably placed on the side of the canal away from the towpath. This was yet another instance of providing an unobstructed course for the towlines of horse-drawn vessels—a point no doubt overlooked in their annoyance by today's holiday makers who find that to open a bridge they have to jump from boat to towpath, run along it, cross over the bridge to open it, and then cross back again afterwards.

Piecemeal destruction of bridges, and their replacement by culverts or low-level fixed bridges without headroom for boats, usually followed quickly on closure of a canal to navigation. Opening bridges were prime targets, being fixed in the closed position or replaced by stronger structures. Sometimes they were fixed before closure. Lifford Lane swing bridge on the northern Stratford Canal was a notorious example: damaged by a lorry during the Second World War, it was replaced by a low-level bridge which could be opened only by jacking it up after twenty-four hours' notice. At the instance of the infant IWA, a new swing bridge was installed in 1949. On the Pocklington Canal, too, swing bridges were replaced by fixed low-level bridges in 1962 while it was still a statutory navigation: some have since been replaced by swing spans as part of the restoration scheme for that canal.

Some of the opening bridges on the Forth & Clyde have been replaced by low-level bridges or culverts; and a concrete bridge which carries a road now crosses the chamber of lock 20, Wyndford, marking the site of a former bascule-bridge.

Fixed bridges too, which were adequate for eighteenth century road traffic, became an obstruction to traffic of the twentieth century. In many places an improved road, straight and easily graded, crosses a closed canal by a culvert which marks the site of a former bridge. There are examples on the Shropshire Union Canal, the Union Canal and elsewhere.

Junctions

There is a particular fascination about a canal junction. After cruising for hours along a canal which is its own complete and private world, one receives an all too brief glimpse of another canal, framed by the arch of a roving bridge. It serves as a reminder that one's own route is linked with the wider world in its own way—but however inviting the other canal may appear, there never seems to be time to explore it immediately. It encourages a return visit.

Like other canal features, canal junctions are revealing about the past. Generally they were laid out in T form, with the waterway at the actual junction made wider than usual, for boats to manoeuvre. Some junctions built late, however, were laid out to suit prevailing traffic. Boats leaving the Staffs & Worcs for the Birmingham & Liverpool Junction Canal at Autherley have a gentle turn if approaching from the south, that is, the Birmingham direction, but an awkwardly sharp one if approaching from the north. Similarly, at the far end of the B & LJ, the earlier Chester Canal approached its terminal basin at Nantwich by a curve: the B & LJ was laid out to continue straight on at the start of the curve. This left boats coming from the South and attempting to enter the basin with a very sharp turn indeed. I know from experience! Almost as sharp, if approached from the north, is the turn from the Trent & Mersey on to the short arm leading to the Anderton Lift: it was laid out for traffic going to and from Middlewich and the Potteries, both south of Anderton.

Where the new Birmingham main line diverged from the old, junctions were laid out to give boats on the new line a straight run, or one that was very easily curved. The junction called Braunston Turn was laid out as a triangle with concave sides and a central island. Here one of the Oxford Canal's cut-offs diverged from the original line, but part of the original line which was being bypassed was retained to connect with the Grand Junction Canal. Busy traffic was likely along all three sides of the triangle.

The signposts which appear at canal junctions, though often fascinating, are modern. Working boatmen did not need them, knowing their way from a

lifetime's experience, and in any case would have been unable to read signposts if provided.

Some junctions have unusual layouts which are explained by their history. Boats coming south down the Coventry Canal have to make an awkward U-turn to enter the Oxford Canal at Hawkesbury Junction. Originally, the two canals ran parallel for some distance to a junction nearer Coventry and the present junction, however inconvenient, is an improvement on the original. The layout of the junction of the Shropshire Union Canal with its Dee branch at Chester is also most curious. Going north, the main canal descends Northgate staircase and then turns sharp right under a bridge. The Dee branch can then be seen running parallel and rising through two locks to join the main line, which continues in the direction of Ellesmere Port. Going back down the Dee branch, it too makes a sharp right hand turn before locking down into the river: and from a viewpoint at the top of Northgate Locks, it can be seen that the lower part of the branch is in a straight line with the main line staircase.

The original canal here was the Chester Canal which was no doubt laid out in a straight line from the top of the staircase to the river. Later, the line to Ellesmere Port was built—the Wirral Line of the Ellesmere Canal—and then or later still the line from the Dee must have been diverted to join it at the present junction, and a short part of the original canal filled in. Precisely when or why this work was done I have yet to establish (any reader who can elucidate is welcome to write to me c/o the publisher!).

Traces of lost junctions, where an abandoned canal joins one which is still in existence, survive in many places. Where the closed canal has been filled in, the junction is sometimes marked only by a short length of bank piling of different pattern or alignment to the rest—as where the Union Canal formerly joined the Forth & Clyde above the latter's Lock 16. More noticeable are those places where a short length of closed canal remains in water, as far as the first convenient place to stop it off—say its first lock, as at Frankton Junction on the Shropshire Union where the closed line to Newtown diverges from the open line to Llangollen.

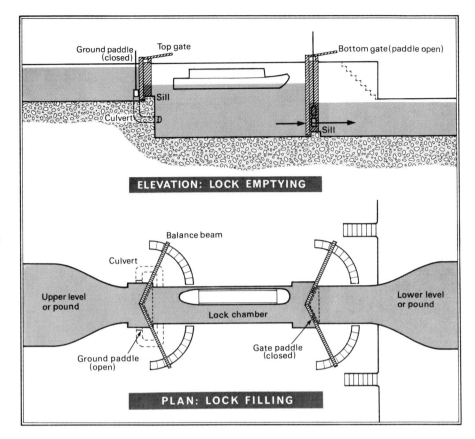

Ground paddle (closed) Top gate Bottom gate (paddle open)
Sill Culvert Sill

ELEVATION: LOCK EMPTYING

Balance beam
Culvert
Upper level or pound Lock chamber Lower level or pound
Ground paddle (open) Gate paddle (closed)

PLAN: LOCK FILLING

7/18 Principal components of a typical lock.

In some places that first length of closed canal is still in use as moorings—as on the Stroudwater Navigation at Saul Junction, where it joins the Gloucester & Sharpness, or at the Wey & Arun's junction with the River Wey. Norbury Junction on the Shropshire Union still looks like a junction but the branch which leads off under the roving bridge, which once went to Shrewsbury, terminates in a few yards at a hire cruiser base.

Great Northern Basin, Langley Mill, which now appears as a basin of rather strange shape at the head of the Erewash Canal, is another lost junction which contains traces of two other canals. Originally, going north, the Cromford Canal diverged from the Erewash where it approached its terminal basin, and continued the line northwards. Above its first lock it was in turn joined by the Nottingham Canal which, in competition with the Erewash, had been running roughly parallel to it up the same valley. Both the Cromford and the Nottingham Canals were eventually closed and, in the vicinity of Langley Mill, filled in; the Erewash Canal's Langley Mill Basin was filled in too after the upper part of that canal was closed in 1962.

The rest of the Erewash Canal was retained for water supply, however, and, to prevent its becoming derelict like other canals in the district, the Erewash Canal Preservation and Development Association was formed. This campaigned successfully for its restoration for navigation. Then, to replace the lost basin at Langley Mill, it restored by its own efforts the derelict bottom lock of the Cromford Canal and the derelict area which was formerly the Cromford Canal/Nottingham Canal junction. It is this which is now called Great Northern Basin. Subsequently, to enlarge the area of water, some members of the association formed a company which has re-excavated and brought back into use part of the filled-in Cromford Canal.

Locks

On river navigations where water was plentiful, and available naturally, locks could and did have chambers made with sides of wood or sloping turf, or a combination of turf and brickwork. It did not matter, within

reason, if the lock chamber leaked, and locks of the latter type survive on the Kennet Navigation. On artificial canals, however, and particularly on those which crossed watersheds, where water was never to be squandered, it was essential for lock chambers to be as watertight as possible, and this meant construction of brick or stone. So far as I am aware there were very few exceptions to this during the canal era. One was at Beeston on the Chester Canal where Telford rebuilt a lock with a chamber of cast iron to counteract trouble from unstable foundations. There were also a few instances where parts of lock chambers are cut from solid rock—as, for example, at Northgate Locks on the same canal. Some recent locks have been built of concrete.

Generally speaking there has been little development in the principles of locks, and the way in which they work, from Brindley's time until the present day; although between one locality and another there were and are many variations in detail, and also in nomenclature. The main components of a typical lock are identified, and the principles on which it operates are shown, in the illustration no. 7/18.

I certainly do not intend to attempt to describe here in writing how to operate a lock, for it is almost impossible to do so simply, without getting over-involved in detail. In any event, generations of illiterate boatmen understood locks, and worked them with ease.

The diagrams are of course simplified. In many locks there are gate paddles in the top gates, as well as the bottom gates. Paddles themselves are raised or lowered, to allow water to flow (or not) from upper pound to lock chamber and from lock chamber to lower pound, by paddle gear mounted above them on gate or ground. The gearing traditionally comprises a rack attached by a rod to the paddle, and a pinion to raise the rack: this is made to turn by a cranked spanner called a windlass, or lock key, fitted on a square on the end of its shaft. In some cases additional reduction gears lighten the task. A ratchet or catch prevents the paddle's descending under its own weight. Working locks on canals is generally the responsibility of boat crews, who carry their own windlasses.

There was much variation from

7/19 Napton bottom lock on the Oxford Canal is a typical narrow lock; less typical is use of lower gates made of cast iron.

canal to canal in the details and positioning of paddle gear, but little development of it during the canal era. In recent years paddles on some commercial waterways have been mechanised, and a hydraulic paddle gear (operated by windlass) has been introduced on cruising canals. It has met a mixed reception. The traditional types of paddle gear are still commonest.

Wide locks have two gates top and bottom. Narrow locks have a single gate at the top end and, usually, double gates at the bottom end. On a few narrow canals such as the southern part of the Oxford single bottom gates were fitted and continue to be used: they were cheaper to build but harder to work. The material is generally wood, although in a few instances cast iron was used (the Oxford Canal is again an example) and in recent years many lock gates have been made from steel. Gates are opened and closed by leaning against the end of the balance beam by the small of the back. On the Caledonian Canal they were formerly opened and closed by chains from

capstans: the capstans remain though the gates have been mechanised, as have the gates on other commercial waterways.

In some narrow locks, particularly on the Trent & Mersey, an effect of opening the ground paddles at the top of the lock is that the inrush of water flows along the bottom of the lock chamber and produces a strong *forward* eddy at the surface; the same effect can be experienced below the bottom gates when a full lock is being emptied to make it ready for an ascending boat. It is my opinion that this feature was designed into locks intentionally to hold the bows of unpowered horse boats steady against the gates. Brindley and his assistants were extremely knowledgeable about the behaviour of flows of water in confined spaces. It was particularly important to hold a boat forward when it was rising in a lock, lest the rudder be damaged against the lower

gates: the bows were protected by a fender, and the sill and top gates by rubbing plates.

Wooden bollards beside many locks are today used by holiday-makers to hold their boats, short and light compared with trading craft, in position while the lock is being filled. But they are often inconveniently placed for this, too near the gates, and it was not their original function. Generally they were placed so that a line could be passed round them to slow and then stop a heavy-laden but unpowered boat entering the lock. Present-day motor craft are put astern to slow and stop them, but reliable and instant ability to run astern was not a feature of the engines fitted to some early motor boats and barges and—particularly on the Grand Canal in Ireland—boatmen continued to use stop ropes and bollards even for motor barges.

Dimensions of locks have been

mentioned in earlier chapters: in general, in England, the clear space in the lock chamber available for boats approximates to 72 feet by 7 feet on narrow canals, 72 feet by 14 feet 3 inches on wide canals in the southern part of the system, and 60 feet by 14 feet on wide canals in the North, where they have not been enlarged. A boat which is able to traverse the whole connected waterway system must therefore be built within these dimensions. H. R. de Salis travelled over much of the inland waterways system in the 1890s, before preparing the noted gazeteer *Bradshaw's Canals and Navigable Rivers of England and Wales*, and built his steam launch *Dragon Fly No. 3* to what he considered to be the maximum dimensions for the purpose: 59 feet long by 6 feet 8 inches beam. These proved satisfactory. The only place where he had trouble was on the Yorkshire River Derwent where the boat was such a tight fit lengthwise in the locks that she was brought down them stern first. Presumably the rudder was put hard over against the bottom gates (and kept well clear of the top gate sill) for he subsequently quoted the maximum length for vessels to use this waterway as 55 feet.

The depth of fall of locks—that is to say the vertical distances through which boats are raised or lowered—ranges from 1 to 14 feet; (from this I except stop locks, to which I shall refer shortly). Both these figures are extremes: a fall of 6 to 8 feet is typical. Contrary to what one might at first suppose, there was a tendency for locks built towards the end of the canal era to be smaller and shallower than those built earlier. Several wide locks at the eastern end of the Trent & Mersey, for instance, have falls of about 12 feet, but the narrow locks of the Birmingham & Liverpool Junction are typically a little over 6 feet deep. Small, shallow locks are quick and easy to work and save water.

The tendency was not invariable, however. Locks on the Shropshire Union's Middlewich Branch, a contemporary of the B & LJ, are over 11 feet deep: but since this canal is at a lower altitude it may be that water supplies were considered less critical. There was also a tendency for all the locks on late canals to be of similar depth, while those on earlier canals varied one from another: this caused

waste of water, for the amount of water drawn from a summit pound by a boat ascending to it is the amount required by the deepest lock on the ascent.

This particular feature can be seen at its most acute on the southern Oxford Canal. Aynho Weir Lock (grid reference SP 494337) has a fall of only 1 foot, but the next lock down is Somerton Deep Lock with a fall of 12 feet. Aynho Weir Lock is shallow because, immediately above it, the River Cherwell is let into the canal on one side and out again over a weir on the other, which meant the builders avoided the expense of an aqueduct and obtained a supply of water. It also meant that the level of the pound concerned rose and fell according to the amount of water in the river, and was kept as short as possible; but the lock down from it, lying on the floor of the valley, could have only a shallow fall. So that it would let down sufficient water to work the next and subsequent locks it was built with a wide chamber, which narrows at the ends to gates of normal narrow width. Shipton Weir Lock, further south on the same canal, has a similar shape for a similar reason.

Almost all locks have bypass weirs, set in the canal bank near the top gates, which lead to side channels or culverts running parallel to them and back into the lower pound. Down these, all being well, there is a continuous small flow of water. This compensates for water lost by evaporation, leakage and operation of locks lower down the canal.

The side pond is a device for saving water. An additional paddle at the side of a lock chamber opens or closes a sluice communicating with the side pond itself, which is at a level intermediate between that of upper and lower pounds. When a boat is about to descend a lock, this paddle is raised and the water run from lock to side pond until its levels equalise. Then the paddle is closed and the rest of the water run into the lower pound in the usual way. The water in the side pond is available for use by the next ascending boat. Many locks on the Grand Junction Canal were equipped with two side ponds at different levels.

The term side pond is also used in a slightly different sense: for a lateral extension of a short pound between two locks, which would otherwise not have sufficient capacity for the water required to operate the lower lock.

Such side ponds may be seen between the locks at Devizes, Kennet & Avon Canal and Marple, Peak Forest Canal.

The Regent's Canal was built with pairs of wide locks side by side with intermediate paddles, so that each one of a pair could act as a side pond for the other. In most instances, however, one lock of each pair is now out of use. When the Oxford Canal Company's three narrow locks at Hillmorton, on its nothern section, were duplicated in 1840 to cope with heavy traffic, the same scheme was adopted. The pairs remain in use, or did until recently.

There were other locations where locks were duplicated to cater for heavy traffic in the pre-railway era. The Trent & Mersey west of Harecastle is one of them: most of the locks are in pairs, although today it is not unusual for only one of a pair to be in use. Some of the locks on the Ashton Canal also were duplicated, and traces of the second chambers can be seen. The Grand Junction company duplicated the wide locks at Stoke Bruerne in 1835, and about 1838 added narrow locks alongside many of the wide locks on the climb up the northern slope of the Chiltern Hills to Tring: many of the craft passing at that time were single narrow boats which wasted water when they passed through a wide lock alone. In neither case did the duplicated locks last very long in use, for the railway era soon arrived, but traces of them can still be seen in the form of double-arched bridges at the tails of locks, particularly at Stoke Bruerne where the top lock of the flight was built as the duplicate and the chamber of the original lock alongside now contains open-air exhibits of the Waterways Museum.

Nearer to Birmingham on the Grand Union Canal north of Napton the narrow locks were replaced by wide ones during the 1930s. The new wide locks were built in the dry alongside the old; the narrow locks remained in use for some time but eventually all were either demolished or turned into bypass weirs. Their remains are still prominent.

Instances where locks have been lengthened to take bigger craft are regrettably rare on canals in the British Isles, but can be seen in a few places in Yorkshire. It is sometimes possible to distinguish between the old masonry of the original lock and the newer work of the extension, and gates in the original positions are sometimes retained for passage of small craft. There are examples at Castleford on the Aire & Calder and Doncaster on the Sheffield & South Yorkshire.

At the opposite extreme come the locks on tub-boat canals. None have been used for many years, but on the Shrewsbury Canal several survive, though derelict. It is immediately obvious to the practiced eye that they are both longer and narrower than an ordinary narrow lock: they are rather more than 81 feet long and 6 feet 4 inches wide, and were able to take four tub boats at once. The most striking feature, though, of these locks is the guillotine gates at their lower ends (the top gates are of the usual type). The guillotine gates, when operable, rose and fell vertically within a framework of massive timbers; chains supporting them led over pulleys to counterweights. The latter originally were large wooden boxes filled with stones, which were suspended immediately downstream of the gates, but from the 1840s onwards the gates were modified and counterweights which rose and fell in wells alongside the gates were fitted. The only survivor of the original type is at Hadley Park (grid reference SJ 672132); a gate of the later type can be seen a little way to the south.

Why guillotine gates were fitted at all is something of a mystery: it is said to have been to save water, as by using them instead of ordinary bottom gates the length of the chamber could be reduced by about three feet, that is, the space taken up by ordinary gates when opening and closing. If this was indeed the reason it seems a cumbersome way to achieve a small benefit. I have a theory that the reason may have been this: when passing four loosely-connected, square-ended tub boats through a lock together, one would want to make them as tight a fit, lengthways, as possible, lest turbulence cause damage by battering them successively against one another and the gate or gates. In these circumstances, to eliminate that gap of three feet might well be worthwhile.

Whatever the reason, one thing is clear: guillotine gates (which must rise high enough to clear boats passing

7/20 Stop rope in use on a barge entering a lock on the Grand Canal Circular Line, Dublin.

beneath) were more practicable on a tub-boat canal than an ordinary canal because of the limited headroom required. The headroom on the Shrewsbury Canal was 5 feet 5 inches; few narrow canals had less than 7 feet headroom.

So far as I am aware, the only lock on a narrow canal with guillotine gates is the stop lock at King's Norton Junction, where the Stratford Canal joins the Worcester & Birmingham. Stop locks were built where one canal joined another, to minimise loss of one company's water to the other. Junctions were laid out so that the newer canal was at a higher level than the old, and therefore fed it with water rather than the other way about; but the fall was usually only a few inches, so that the new canal lost no more water than necessary. This meant that in normal operation levels sometimes fluctuated with the level of the new canal lower than the old. The Stratford was laid out with its intended water level two or three inches higher than that of the W & B, and in these circumstances ordinary gates would not work reliably: guillotine gates could operate against a very small head of water, no matter what the relative levels of the two canals were.

Other stop locks in comparable

7/21 Side pond at a lock of Marsworth flight, Grand Union Canal. This saves water: when a boat is descending the lock, water is run into the side pond, where it is stored to be used later by another boat.

situations had two sets of gates, to open in either direction, used according to which canal was the higher. One such stop lock is at Hall Green, where the Macclesfield Canal joined the Trent & Mersey: today the fall is always from the former to the latter, but the lock has two consecutive chambers of which one is in use and the other, though it no longer has gates, can be seen from the positions of the quoins into which the gates once fitted to have been intended for a fall from the Trent & Mersey to the Macclesfield. Another stop lock with provision for gates to open in either direction can be seen on the Duke's Cut, a connection between the Oxford Canal and the upper Thames north of Oxford. Here, the cause is slightly different: formerly, the level of the Thames used to fluctuate so that it was sometimes higher than the canal, sometimes lower.

Stop locks only remain in use where there is still a difference in levels between adjacent canals, as between the Coventry and Oxford Canals at Hawkesbury Junction. Elsewhere, where two adjoining canals have passed under a single ownership and their levels are the same, the stop locks have gone out of use, although their chambers often remain as reminders of the past. There are examples on the Ashby Canal at Marston Junction where it joins the Coventry, on the Coventry at Fradley Junction where it joins the Trent & Mersey, and on the Worcester & Birmingham at Worcester Bar (Gas Street Basin), Birmingham, where it joins the BCN. The name Worcester Bar is a reminder that

originally, so great was the distrust of the Birmingham company for its new rival the W & B, there was a physical barrier between the two canals, a strip of land across which goods had to be transferred from boat to boat. The water connection by the stop lock was made through it later.

Where canals join the sea or the tideway are sea locks, at which the fall varies according to the state of the tide. In places the tide rises high enough for additional gates, pointing outwards, to be required for use when the level of the sea is higher than that of the canal. An example is the complete but unused entrance lock from the tidal Medway to the basin of the Thames & Medway Canal at Frindsbury, Strood.

On the Gloucester & Sharpness Canal, pleasure craft moor at Sharpness in what now appears to be an arm of the canal but is in fact the original line, to the original sea lock, both of which were superseded by the present dock and lock in 1874. Sea locks are often larger than ordinary canal locks, to allow seagoing vessels into a basin or dock above them for transfer of cargoes into barges or boats. Ringsend Basin, where the Circular Line of the Grand Canal terminates in Dublin, was equipped with no less than three locks of various sizes, side by side, down into the tidal River Liffey. The smallest lock, intended for barges, and the largest, for ships, remain in use.

The placid appearance of Clachnaharry Sea Lock, at the north eastern end of the Caledonian Canal, belies an epic story of construction. The sea bed shelves gently here, and so that ships might enter at any state of the tide it was necessary to build the lock 400 yards out from the shore. At the site chosen there was mud to a depth of 55 feet over a hard bottom, mud of such a consistency that oak piles rebounded when attempts were made to drive them in, and it was impossible to build a coffer dam. Telford's solution was to build a great clay embankment out from the shore, and weight it with stone at the site of the lock so that over several months it gradually settled, parting the mud, on to the rock beneath. Only then could a coffer dam be made and the lock, at that period (1804–5) the largest in the world, be built.

So far I have dealt with locks which are complete, or almost so. What happened to locks after a canal was

7/22 Duplicated wide locks at Hampstead Road on the Regent's Canal enabled heavy traffic to be passed, while saving water. Castellated building on the right is an unusual lock cottage.

closed or became disused varied according to circumstances, such as whether the canal was still in water, either naturally or by intent for water supply, and whether or not they were in a populous area. Many locks were just left to decay gently, like those of the Wey & Arun and Shrewsbury Canals. In some happy instances it has been possible to restore such locks and re-use them. Locks on the southern Stratford, Ashton and Kennet & Avon Canals are a few of many examples.

In some instances the locks of closed canals have been deliberately destroyed or filled in in the interest of public safety, for a derelict lock is both an attraction for children and a danger to them. Where a water channel has to be maintained for water supply, they have in places been turned into 'cascades'. I was examining recently the site of Ripon Lock (grid reference SE 324703), the top lock of the Ripon Canal, which is in this state. Where were once the top gates is now a weir; most of the masonry of the lock chamber has been knocked down into it to give gently sloping sides; and the flowing water finds it own way down a jumble of rubble, rubbish and brambles where there was once a lock. The result may be perhaps safer than a derelict lock, but it is still un-

satisfactory and unsightly. Locks have been treated similarly on the Huddersfield and Monmouthshire Canals.

Where it was not necessary to maintain a flow of water, canals and locks have been entirely filled in. It is now difficult if not impossible to trace the course of the locks which formerly connected the Union Canal with the Forth & Clyde: they were filled in during the 1930s. During a visit to the St Helens Canal in 1978 I found part of it was being filled in (*infilled* in the jargon of bureaucracy) with what appeared to be industrial refuse. Winwick Lock (grid reference SJ 594921) was being filled in without, so far as I could see, any attempt to demolish chamber or gates, and the paddle gear of the top gates projected out of the morass. One might suppose that the lock is being preserved for future archaeologists, provided they do not mind what they dig through! See page 91.

On early lateral canals, locks were positioned at irregular intervals dependent on gradients of canal and river. Cross-watershed canals with long contour-following summit pounds brought the possibility of concentrating locks together in flights. Some Brindley canals show evidence of both sorts of layout, with locks at irregular intervals becoming more frequent and culminating in a flight of several locks close together by which the canal gains its summit pound. The ascent of the Oxford Canal up the Cherwell valley, culminating in Claydon flight which leads to the summit, is a good example.

It was convenient for both operation

and maintenance of a canal to build locks in flights with long intervening pounds, and this became an accepted practice.

The most extensive flights of locks eventually constructed were those of twenty nine wide locks at Devizes on the Kennet & Avon and thirty narrow locks below Tardebigge on the Worcester & Birmingham. Other notably extensive flights of locks are at Wigan on the Leeds & Liverpool (twenty three wide locks), Hatton on the Grand Union (twenty one wide locks) and Stourbridge on the canal of the same name (sixteen narrow locks). All the sixteen locks of the Peak Forest Canal are concentrated in one flight at Marple, and all twelve locks of the Macclesfield in one flight at Bosley.

Where canal builders encountered a steep rise or fall in the surface of the ground which they could not avoid, they joined two or more successive locks together in a continuous structure. A single pair of gates separated each lock chamber, and combined the roles of top gates of the lower lock and bottom gates of the upper lock. Such locks are called risers or staircases in Britain and double locks in Ireland, where none were built with more than two chambers. Double locks are quite common on the Grand and Royal Canals, particularly in the course of the climb which each canal has to make out of Dublin. Two-lock staircases were less common in England, though there is a narrow example on the Staffs & Worcs Canal at Botterham and a wide one on the Shropshire Union at Bunbury.

The earliest staircase built in England with more than two chambers was Bingley Five-Rise staircase (grid reference SE 107399) on the Leeds & Liverpool Canal, built in 1774. Because the lock chambers are short, it rises with dramatic steepness. A few hundred yards below it is Bingley Three-Rise staircase, with three chambers. Other three-lock staircases were built at Grindley Brook on the Ellesmere Canal and at Chester on the Chester Canal. The Manchester, Bolton & Bury Canal had two successive staircases of three locks at Prestolee (grid reference SE 752064): though long dry, their masonry remains impressive.

The old Grand Union Canal concentrated its locks, all narrow, in two flights, at Watford, Northants, and Foxton, Leicestershire. The Watford flight comprises, from the top, a single lock, a staircase of four, and two more singles; the Foxton flight, the most ambitious of its kind in England, comprises two successive staircases of five with a short intervening pound in which boats may pass one another. This flight is far surpassed, however, by the staircase of the Caledonian Canal. With chambers of ship canal size, to take vessels 150 feet long by 35 feet beam, there are staircases of four locks at Muirtown (Inverness), five locks at Fort Augustus, and no less than eight locks at Banavie near Fort William.

The advantages of staircases were cheapness of construction and maintenance, compared with single locks. The disadvantages are that they obstruct traffic—on a busy canal one can wait for a long time before ascending a staircase while successive boats descend—and they waste water (unless equipped with side ponds) for when a boat is about to ascend a staircase with water at low level in the lock chambers, all except the lowest have to be filled before it can start.

There were instances, therefore, where staircase locks were replaced by ordinary locks. Early in 1978 I was ascending the second lock of Meaford flight of four ordinary locks on the Trent & Mersey (grid reference SJ 890353). Being then unaware of its history, I was surprised to observe that I was clearly on a section of canal built late in the canal era, unlike most of the T & M. The three lowest locks in the flight ran in a straight line, achieved by cut-and-fill, the lock I was at being in a

cutting which shortly gave way to an embankment. Then, as I moved forward along the embankment, the dry bed of an earlier course could be seen away down to the left, with a culvert over a stream still surviving. On this old course there was originally a staircase. A leg-of-mutton shaped lagoon above the third lock indicated where the new route rejoined the old and the top lock of the flight was an original, the poky brick bridge at its tail, innocent of a towpath, contrasting markedly with the split footbridges at the tails of the other locks.

There is a curious flight of three locks at The Bratch (grid reference SO 867938) on the Staffs & Worcs Canal. At first glance they might be mistaken for a staircase, but on closer examination they are seen to comprise three separate locks, although the bottom gates of the top lock are less than a boat's length away from the top gates of the middle one (which means that the water level in each lock and the intervening short pound has to equalise before a boat can move from one lock to the next), and the same applies between the middle lock and the lowest. Each of the two short pounds between the locks is connected to an extensive side pond.

Since it was on this part of the Staffs & Worcs that the first narrow locks were built, and the next lock but one down the canal is a staircase pair, it has been suggested that Brindley designed The Bratch flight as a means of descending a steep slope, and then developed the idea further to invent the staircase. This may well have been typical of the way in which The Schemer worked, but if it were indeed so in this case he would have been rediscovering what others had earlier invented. It seems unlikely, too, that the Duke of Bridgewater, who must have seen staircases on the Canal du Midi, would not have described fully all of that canal's features to the engineer of his own canal. J. Ian Langford, in *A Towpath Guide to the Staffordshire and Worcestershire Canal*, suggests the interesting alternative that The Bratch flight was originally built as a staircase—and altered later to three ordinary locks to save water. This certainly seems likely, although it raises the further question of why the staircase was not simply converted to operate through side ponds like the Foxton staircase locks.

Vertical and inclined plane lifts

The inconvenience of both staircases and extensive flights of locks, where a canal had to ascend or descend through any great difference in level, gave early canal engineers a powerful incentive to design other devices to achieve the same object. These usually took the form of lifts, both vertical lifts, by which boats were raised and lowered vertically, usually in tanks of water, and inclined plane lifts, by which boats were raised and lowered up or down a slope, usually on cradles running on rails. They were commonest on tub-boat canals. Unfortunately the technology of the early canal era was insufficiently far advanced, and at first the vertical lifts and inclined planes that were built often failed to work reliably. Inclined planes, after Ducart's early fiascos in Ireland mentioned in chapter two, generally proved more reliable than vertical lifts.

Tardebigge top lock, on the Worcester & Birmingham Canal (grid reference SO 995693) is at fourteen feet probably the deepest of narrow locks: it is so because it occupies the site of a vertical lift, a prototype, which proved unsatisfactory, of those by which the company hoped to avoid the need to build its innumerable locks down to the Severn. Lifts and inclined planes were especially popular on the canals of the West Country in England, such as the Grand Western's extension to Taunton which had one plane and seven vertical lifts, and the Bude Canal which had six inclined planes. These all seem to have worked well enough. Traces of these lifts, and others in the West Country, can still be seen although all are long-disused; but not having yet had the opportunity to seek them out for myself I am unable to give practical guidance. For those wishing to know more, Charles Hadfield's *The Canals of South West England* would be as good a starting point as any.

The other area where inclined planes were used in quantity was East Shropshire, where the tub-boat canals had six. One of these, at The Hay on the Shropshire Canal, was last used about 1894 but now forms part of the Blists Hill section of the Ironbridge Gorge

7/23 Remains of tub-boat locks on the Shrewsbury canal at Hadley Park, (grid reference SJ 674131). That the lock chamber is slightly narrower than usual is clear to the experienced eye – the boats were 6 feet 4 inches wide. The guillotine gate was raised and lowered in its frame by the winch (right); a counterweight fell and rose in a pit beside the ground level pulley, left.

Museum, and has been partly restored. Railway track has been re-laid upon it—in their latter days these planes were laid with ordinary railway track—and at the top of the incline the masonry of the short reverse slope, by which the cradles descended into the upper canal, has also been restored, and there are the remains of supports for the winding machinery and the mountings for steam engine and boilers. Generally this inclined plane, which has double track, was operated by allowing the weight of a descending laden boat to haul up an empty ascending one: but a steam engine was needed to haul boats up out of the water of the upper canal, and also when a heavier load had to ascend the plane than was descending. At the bottom of the plane, the canal which was dry for many years is now once again in water, and the rails disappear beneath its surface.

Addressing the Motor Yacht Club in February 1908, H. R. de Salis was able to tell them that the Foxton Inclined Plane Lift on the Grand Junction Canal would pass barges measuring 72 feet by 14 feet, and had 'superseded the flight of ten locks formerly in use at that place'. The story of Foxton Inclined Plane (grid reference SP 693896) is perhaps the most evocative of canal lost causes. It had been brought into use in 1900, yet later in the year that de Salis spoke the locks, which are narrow, were brought back into use part-time, and in 1910 regular use of the plane was discontinued. Its origin, in the purchase of the canals between Norton and Leicester by the Grand Junction company in 1893 and that company's, and Fellows Morton & Clayton Ltd's, desire to regain the Derbyshire coal traffic, has been told in chapter six.

Foxton was laid out so that boats were carried in two large tanks of water which counterbalanced one another and travelled sideways up and down the plane on rails. Power was provided by a steam engine. Hydraulically-operated guillotine gates closed the ends of the tanks and hydraulic rams pressed the tanks sideways when necessary to seal them against the upper level of the canal. At the lower level, the tanks were submerged. The vertical rise was 75 feet, the gradient 1 in 4, and the passage through the lift took about 12 minutes, compared with over an hour through the locks. After teething troubles, the plane seems to have worked satisfactorily.

It was lack of sufficient traffic that made operating the plane uneconomic, for it needed a staff of three and the boiler had to be kept in steam, needing constant attention. Had it been built a few years later when electric power was available, things might have been different. As it was, after closure, the lift gradually became derelict; the machinery was sold for breaking up in 1928, and the resultant scrap carried away by boat. The site then became very overgrown, until in the early 1970s the top part of it was cleared and made accessible to the public. It was scheduled as an ancient monument in 1973. The bottom basin remains in water and is used as moorings for pleasure craft.

The brief life of Foxton Inclined Plane contrasts markedly with that of the only comparable canal structure in Britain, the Anderton Lift (grid reference SJ647752). This antedated the Foxton Lift by twenty five years and happily is still in use. It operates vertically and provides the only direct link between the River Weaver and the small canals.

The difference in water level between the Trent & Mersey Canal, at the top of the lift, and the River Weaver is normally 50 feet 4 inches. As completed in 1875, the lift worked hydraulically. It had two wrought iron troughs of water, each large enough to take one barge or two narrow boats, and each of these was supported by a large hydraulic ram, the presses of which could be made to interconnect by opening a valve. The bulk of the lift was achieved by syphoning water out of the lower trough to reduce its depth by six inches, and then allowing the upper trough to descend by gravity, raising the lower trough at the same time. When the descending trough entered the water of the river it could no longer raise the ascending trough which was then pumped by hydraulic pressure to the top of its stroke. At this point its water level was six inches below that of the canal, and canal water was run into it to replace that syphoned out earlier. Hydraulic pressure was maintained by a steam engine.

On the whole this worked well until the turn of the century, but trouble was experienced in the rams from electrolytic corrosion, and when the steam boilers wore out in 1902 an electric pumping plant was installed to provide hydraulic pressure. This took power from a local supply which was already available. By then, however, it was becoming more and more difficult to keep the rams sound, the hydraulic pipes watertight and the structure safe. In 1906 it was decided to reconstruct the lift.

The entire operation was now to be electrically powered, with the troughs worked independently and counterbalanced by weights suspended by cables over overhead pulleys. Since the load was now to be taken from above, the meant that a new foundation for the lift had to be built, and a great many more supports provided. The lift in its reconstructed form came into use in 1908.

This is the form in which it still operates. Extensive structural repairs in 1974–5 included grit-blasting much of the structure down to bare metal, replacement of corroded girders, and repainting.

Highest pounds

Having considered both canals on a level and the means of changing levels, it seems appropriate to mention here that the highest altitude of any canal in the British Isles is 645 feet above sea level, attained by the summit pound of the Huddersfield Canal which includes Standedge Tunnel. This is not now navigable; the highest pound which is navigable is that comprising the highest levels of the Macclesfield and Peak Forest Canals from the top of Bosley Locks to Marple and on to Whaley Bridge. This is 518 feet above sea level and heather from higher hills to the east spreads over the sides of cuttings near Macclesfield.

7/24 (right) The Anderton Lift in the early 1960s, with a pair of British Waterways Admiralty Class narrow boats approaching from the River Weaver.

CHAPTER 8

FEATURES OF CANALS: BOATS AND BARGES

Survivors among trading craft

I have already suggested the horse as that canal feature of which the absence is the most conspicuous: but the trading boat, on small canals, runs it a very close second. In some respects its absence is even more noticeable, for its disappearance is more recent, and reminders of its former presence more obvious.

On some commercial waterways, trading craft are of course still very much present. Some of them are of modern design, such as the push-tows of two or three dumb barges propelled by a pusher tug on the Sheffield & South Yorkshire and Aire & Calder Navigations; but others, such as the compartment boats of the same waterways, known as 'Tom Puddings', are old enough to be worthy of description in this chapter.

On cruising canals there is still some trade, often by carriers who are enthusiasts for water transport and use traditional craft. There are also many traditional craft, formerly used for carrying, which are now used by BWB and other navigation authorities to carry materials for maintenance. Many other former trading craft have been converted for pleasure use. The most basic manifestation of this, the camping narrow boat, is scarcely altered from trading condition, for the campers sleep and cook in the hold under tarpaulin cloths altered from tradition only by insertion of clear patches to let through some light. Some such boats revert in winter to carrying.

Other narrow boats have been converted to carry passengers: both for short outings, in which case the hold is fitted with seats, and some sort of awning protection is provided against rain; and for longer holidays, in which case the hold is built over as a long cabin and the boat becomes a floating, mobile hotel. Conversion of boats in this way for private use commenced with a few in the 1920s and 1930s, and became popular after the Second World War. Then, because a full-length narrow boat is difficult for an inexperienced amateur to steer, some were shortened before conversion. From these evolved the steel-hulled canal cruisers of today, built new in large quantities for both private and hire use, generally between 20 and 60 feet long, narrow beam, and with hull form based on the traditional narrow boat.

Examples of many types of trading craft formerly used on canals have been preserved—by individuals, by groups set up for the purpose, and by museums. Some of these are on public view, others are not. Some such boats have been restored to the colour schemes of former owners so that, for instance, the inscription *Fellows Morton & Clayton Ltd* is seen again on canals; and some boats converted for pleasure cruising have later been reconverted back to trading condition.

Finally there are the derelicts, gently disintegrating in the mud of backwaters and disused arms.

Boats other than derelicts generally being mobile things. it is not simple to suggest locations where complete boats of any particular type may reliably be seen. But examples of traditional craft formerly used for trade are likely to be seen on most of the appropriate canals during the course of a cruise along them, and sometimes elsewhere—such as the narrow boats used for maintenance on the River Thames. Where boats are on public view in museums I mention this below; privately preserved boats can often be seen gathered together at rallies of boats, particularly the Inland Waterways Association's annual National Rally.

Materials and means of propulsion

Before considering the individual types of boats and barges, it is worth saying something, briefly, about the materials from which they were made, and then, more extensively, about the means used to propel them, for the permutations of these three factors are many.

Originally, canal boats were built of wood—mainly oak and elm—and wooden boats and barges for canal carrying were built until as recently as the mid-1950s. Their builders then were still using the techniques, designs and materials of much earlier days, and since some of these vessels have been preserved, the results of using such ancient techniques can fortunately still be seen in boats and barges that are still fairly new.

The first iron boat was launched on to the Severn at Coalbrookdale in 1787; but it was not until 1811 that the first iron boats specifically intended for canal use were built, when iron tub boats were introduced on to the Tavistock Canal. The Forth & Clyde Canal's passenger packet boat *Vulcan*, launched in 1819, was the first iron boat built in Scotland (and so the precursor of a great industry). Later, boats of iron, and then steel, became common, but they had little clear cut advantage over wood. Wooden hulls were more elastic than steel, and repair yards for wooden boats were for many years more common and easy to find in event of damage. On the other hand, wood was quickly cut by ice, and tended to become sodden with age, but then both iron and steel corroded. A compromise was the 'composite' narrow boat, with iron sides and elm bottom.

On river navigations, barges were

8/1 Sail on a canal: a Humber keel on the Fossdyke.

originally either hauled by gangs of men ('bow hauling') or sailed, when the wind was favourable. Canals generally were built for horse haulage, the Stroudwater Navigation, so far as I can be certain, being the only canal originally intended for bow-hauling throughout. But human propulsion survived in modern times over short lengths of canal, particularly where the butty of a motor narrow boat had to be bow hauled through a flight of locks, and in the form of legging or poling unpowered boats through tunnels without towpaths. Members of the Dudley Canal Trust demonstrate both the latter techniques from time to time.

Sailing on canals by commercial craft was unusual, because of the narrow, winding channels of many canals and their frequent bridges; but Mersey flats used to sail on the St Helens Canal in the early days, and Humber keels sailed along the Fossdyke until well into the present century. Fishing boats traversing the Caledonian Canal also used their sails.

So the principal means of propulsion for canal boats and barges, for over 150 years, was the horse, and, to a lesser extent, the mule and donkey—the latter working in pairs. Horse haulage continued to be significant into the 1950s and to this day carrier Caggy Stevens continues to keep one horse to move day boats laden with rubbish about those Birmingham canals where flights of locks would give a tug no advantage. Otherwise horse haulage is limited to passenger tripping boats. One of these operations, at Llangollen, is very old established: it started in 1884, and has continued ever since, except, perhaps, during the war years. It employs several boats and several horses. Elsewhere, horse-drawn passenger boats are modern innovations, most of them day trip boats although there is at least one horse-drawn hotel boat. Many of the boats are converted trading craft. These operations tend to have an ephemeral quality, dependent upon the whim of the operator, but places which have or have recently had horse-drawn trip boats include Chester and Norbury Junction on the Shropshire Union Canal, Newbury on the Kennet & Avon, Tiverton on the Grand Western, Cromford on the Cromford Canal, Robertstown on the Grand Canal, and the Peak Forest and Ashton Canals.

The early development of the application of steam power to the propulsion of boats and ships took place on inland waterways, much of it on canals. In 1788 William Symington, under the patronage of Patrick Miller, fitted a small double-hulled boat with a steam engine and tried it out on the latter's loch at Dalswinton, Dumfriesshire. The reciprocating motion of the engine was converted to rotary motion at the paddle wheels by a system of chains and ratchets, and the boat worked well enough for a full-size version to be built and tried out the following year on the Forth & Clyde Canal (of which Miller was one of the proprietors). This too worked well, but not well enough for its patron. There was also a question of whether Symington's engine contravened patents held by James Watt; at any rate, Miller withdrew and it was not until after Watt's patents expired in 1800 that Symington's next vessel was built.

This time his patron was Thomas Lord Dundas, Governor (i.e., chairman, in today's terms) of the Forth & Clyde company, and the vessel was *Charlotte Dundas*; drive was by connecting rod and crankshaft to a single paddle wheel on the centre line of the hull and the vessel, intended as a tug to replace horses, was technically wholly successful. She demonstrated her powers when, in March 1803, she towed two laden vessels, each of 70 tons capacity, for $19\frac{1}{2}$ miles along the canal's summit pound in 6 hours—and this on a day when, as Symington recorded, 'it blew so strong a breeze right ahead . . . that no other vessel on the canal attempted to move to windward.'

Nevertheless *Charlotte Dundas* was never put into regular service, ostensibly because of fears that her wash would damage the canal banks. These fears were justifiable, but another factor was probably that Dundas had his enemies among the other canal proprietors and could not get their approval.

Through Lord Dundas, Symington obtained an introduction to the Duke of Bridgewater. The Duke had already experimented with a steam paddle tug on his canal in the late 1790s: this too had been successful, but not sufficiently so, and he was at first unwilling to consider the subject further. But having been shown a model of *Charlotte Dundas*, he reversed his opinion and ordered Symington to build eight full-sized vessels. Then, before work could start on them, the Duke was dead and the order was countermanded by his successors.

It was left therefore to Henry Bell, who had witnessed Symington's experiments of 1789 and 1803, to build the first steamship to enter commercial service in Europe—the *Comet*, in 1812. This worked principally on the Firth of Clyde, but visited the Forth in 1813 and so must have passed through the F & C Canal; later she also worked on the Crinan Canal.

Relics of these developments are few. Part of the rudder of the *Charlotte Dundas* survives in Grangemouth Museum (Bo'ness Road, Grangemouth), the principal exhibit in a display relating to her, and the engine of the *Comet* is in the Science Museum, London, as is that of Symington's 1788 Dalswinton boat.

The success of the *Comet* started a minor boom in steamer services on estuaries, rivers and coastal waters. The Crinan and Caledonian Canals got their share in this, but otherwise canals tended to be passed by, although there was a second generation of more-or-less experimental paddle steamers on the Forth & Clyde in the 1830s and steam paddle tugs started work on the Aire & Calder in 1831. Then, early in 1837, Francis Pettit Smith carried out trials on the Paddington branch of the Grand Junction Canal with a steam

launch fitted with his patented screw propeller. There had been many earlier attempts to make a screw propeller, but this one was successful and led to the gradual introduction of the propeller in place of paddle wheels. This took place mainly at sea, but was also important for canals, where width of vessels was restricted, and the next forty years saw the gradual introduction on canals of screw-driven steam barges, boats and tugs.

With few exceptions, however, steam barges and boats suffered the disadvantage that steam plant was bulky in relation to the size of the vessels concerned, and needed constant attention, while tugs meant extra work at locks. So not only did steam not wholly replace the horse, but canal carriers were in due course very much quicker to adopt the internal combustion engine than they had the steam engine. Various internal combustion engines were tried experimentally in the 1900s—the Sheffield & South Yorkshire had a motor barge in 1906, for instance, and the engine which was to become almost synonymous with canal motor boat propulsion was first used for this purpose in 1911: the Bolinder.

The Bolinder engine, made in Sweden, was a crude-oil semi-diesel or hot bulb engine. That is to say, the fuel was ignited by the effects of the heat of low compression (about 200 lb/sq. in.) combined with that of an uncooled part of the cylinder head, called the hot bulb. This had to be pre-heated by a blowlamp before starting the engine, but was subsequently kept hot by the successive explosions. The engine was also direct reversing, without a gearbox: direction of rotation of the engine was reversed by an arrangement of rods, tappets and pumps which timed injection of a charge of fuel into the cylinder so that it fired shortly before the piston was due to reach top centre, and so forced it back down again. In the form used on canals, Bolinder engines had single cylinders. An engine of this type, built as late as 1928, is displayed in the Waterways Museum, Stoke Bruerne, and several preserved craft are still powered by them.

British representatives for Bolinder were James Pollock Sons & Co. Ltd, consulting engineers and naval architects; vessels they designed were at the relevant period built by sub-contractors, though later they set up

their own shipyard. Their first customers for canal applications of Bolinder engines came in 1910: Cadbury Bros Ltd and the Grand Canal Company. Both had strong incentives to develop a self-propelled canal boat with a compact engine installation.

That of Cadbury Bros Ltd had arisen out of the progress of its business. Its main factory was at Bournville alongside the Worcester & Birmingham Canal, and since 1905 Cad-

bury had been successfully countering Swiss competition in manufacture of milk chocolate with the then new brand *Cadbury's Dairy Milk*. This success had led it to build an out-factory at Knighton, Shropshire, alongside the Shropshire Union Canal in a dairying district, to collect milk, condense it and combine it with chocolate to produce chocolate crumb, which is partly-processed milk chocolate that keeps well. This had then to be con-

veyed by canal to Bournville to be converted into the finished chocolate. The new factory was opened in 1911.

To work between, principally, Knighton and Bournville, Cadbury ordered two motor canal boats from Pollock late in 1910. These were of narrow beam but far from traditional narrow-boat design—of this, more shortly. They were named *Bournville I* and *Bournville II*, built with 15 bhp Bolinder engines and delivered in

8/2 Haulage of narrow boats by horse, and, in the background, a pair of donkeys.

141

1911: the first canal boats built with engines of that make. The first boat, *Bournville I*, entered service in the early summer.

The Grand Canal Co. had been towing barges in trains by steam tugs since the 1860s—intermittently over its whole line from Dublin to the Shannon, which meant delays at locks, or over the long lock-free levels in the centre of the canal (with horse haulage at the ends) which meant other delays while the barge trains were assembled and broken up: so that tugs brought little benefit.

In 1910 it tried out a De Dion oil engine installed in a horse boat with little success, for there was a severe fire on board. Then, in May 1911 it ordered from James Pollock six Bolinder engines (subsequently reduced to four) for installation in horse boats, and two complete motor barges also with Bolinder engines. Contact was first made with Pollock towards the end of the previous year, for the first numbers relating to the Grand Canal Co. in Pollock's numerical list of ships immediately precede those of the two Cadbury boats. The first converted horse boat was tried out at the end of July 1911 and considered very satisfactory. The two motor barges were respectively named *Athy* and numbered 9M. They were built during 1911–12 and were identical except that *Athy* had a 20 bhp engine against 9M's 15 bhp, and an additional 12-inch high bulwark all round was quoted for. These features suggest that 9M was intended for use on the canal proper and *Athy* (despite her name suggestive of the Barrow) for the river and lakes of the Shannon Navigation over which the G.C.Co. operated carrying services. Certainly by the 1930s she was being used on the Shannon, towing other vessels.

In 1912 another famous name adopted the Bolinder engine, when Fellows, Morton & Clayton Ltd, who already had a fleet of steam narrow boats, built the *Linda* with a 15 bhp Bolinder. Like

8/3 Steam on a canal: one of Fellows, Morton & Clayton Ltd's steam narrow boats. Vulcan was built in 1906 with a gas engine, which proved unsatisfactory; she went into service as a steamer in 1910, and was eventually converted to motor in 1927.

the Irish boats, she was the first of many.

From these beginnings the Bolinder-engined boat became common on the canals of Britain and Ireland. Bolinders did not, however, have things all their own way: other makes of engine, such as Petter, and Widdop, were also popular, and the wholly-diesel engine later came into general use.

Electric traction was little used on British canals, the only lasting example, so far as I am aware, being the Harecastle Tunnel tug which operated between 1914 and 1954. Power was at first drawn from batteries, later from tramway-type overhead wiring. The Staffordshire & Worcestershire Canal between Kidderminster and Caldwell Lock was the scene, about 1924, of experiments with an electrically-propelled boat which drew its power from overhead wires, trolleybus-style, but, though claimed to be successful, the system was not extended. According to Langford's *Towpath Guide*, bases of the poles which supported the wires can still be found along the towpath. The Dudley Canal Trust's battery-electric trip boat has a modern engine installation.

River and estuary craft

Many early canals were built to dimensions suitable for the vessels which were then in use on the river and estuary navigations with which they connected. On these, traditional craft of various types had evolved in isolation from one another.

On the Mersey, and nearby rivers, this meant the *flat*, a sailing barge which had been developing for many years before the Duke of Bridgewater built his canal to appropriate dimensions. On canals, flats were occasionally sailed in the early days, but were generally hauled by horses or, later, tugs; in due course, many flats built primarily for use on canals were built as dumb barges without sails or engines and, when they had to go down the Mersey to Liverpool docks, were towed by a tug, as many as sixteen at a time.

The Mersey flat surviving in best condition—and possibly the only complete survivor—is *Mossdale* at the Boat Museum, Ellesmere Port. She has a timber hull, with rounded bows, could carry sixty four tons, and was built probably in the 1870s. Her original owner was the Shropshire Union Railway & Canal Co., which used her between Liverpool and Ellesmere Port, where her cargoes were usually transferred to narrow boats for onward carriage to the Midlands, although she could go herself as far as Chester or even Nantwich when necessary. Her tar-black hull is surmounted by red, white and blue upper works.

At the same location can be seen a representative of the last generation of working craft built for the canals of the North West, the steel-hulled dumb barge *Bigmere*, built in 1948 to the order of the Manchester Ship Canal Co. for use on the Bridgewater Canal. She was used until 1974 and still, when I looked over her in 1978, had a workaday appearance.

The farthest inland that flats worked was over the Rochdale Canal and, although this canal has not carried commercial traffic for many years, a line of gently disintegrating wooden skeletons of Rochdale Canal flats could still be seen in the pound above its entrance lock in Manchester when I cruised that way in 1978.

8/4 The Bolinder semi-diesel engine, complete with direct-reversing rods and levers prominent in the centre of the picture and built-in blowlamp for starting. This is a 30 bhp version, installed in 1928 in the tug Worcester, *now a Boat Museum exhibit.*

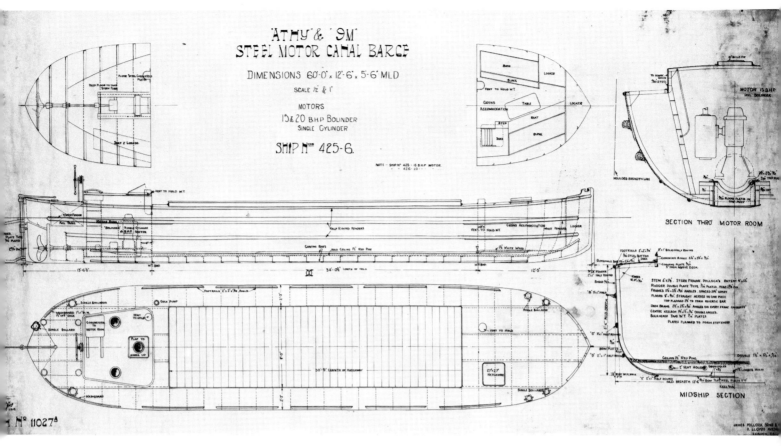

8/5 (above) The Grand Canal Company adopted the Bolinder engine as early as 1911. Many of its early motor boats were converted horse boats, but that same year James Pollock Sons & Co. Ltd, British representatives for Bolinder, designed two motor barges for the G.C.C. which were built during 1911-12; this is one of the drawings. The barge 9M had a 15 bhp engine and its sister, Athy, a 20 bhp engine.

8/6 (overleaf) At the Boat Museum, Ellesmere Port: left to right, Mersey flat Mossdale, narrow boat Chiltern, FMC steam narrow boat Monarch (restored externally); in the right background is the Calder & Hebble motor keel Ethel.

On the Humber and the waterways radiating from it the traditional sailing barge was the Humber keel. The design was, almost certainly, directly descended from that of coasting ships of the Middle Ages and may go back further, to Viking and Saxon ships. The keel's most distinctive feature was its square rig, a single mast carrying main and top sails. Bonthron encountered four of them on the Fossdyke in the 1900s, close together and with all canvas set, and described the sight as 'like the approach of a miniature Armada'. The rig was retained as long as keels traded under sail, which was until 1949, not from conservatism but for convenience in restricted waters. The Humber sloop, with fore-and-aft rig and similar hull, evolved out of the keel in the eighteenth century, and was used in open water: but keel rig was retained for rivers and canals, for a keel with its square sails and lee boards could sail very close to the wind and had no projecting boom to catch waterside obstructions. On canals they traded under sail to Lincoln, by the Fossdyke, and Doncaster over the Sheffield & South Yorkshire.

For trips further inland, masts and sails were removed and stored for collection on the return journey, and keels were towed by horses or tugs. Later, many were fitted with internal combustion engines. Although originally locks on the waterways of the region were designed to fit keels already in use, in later years, as some waterways enlarged their locks and others did not, keels were built to sizes suitable for waterways over which they were to work. The smallest still in regular commercial use are those which, at the time of writing, continue to carry on colliery-to-power-station coal traffic on the Calder & Hebble, and to pass through its locks, measure only 57 feet 6 inches long and 14 feet beam. Not much larger are the 'Sheffield-size' keels which measure 61 feet 6 inches by 15 feet 6 inches and continue to trade on the Sheffield & South Yorkshire Navigation pending its long-delayed enlargement. The permissible draft on this waterway is 6 feet: this, and bows which are so blunt as to be almost square, enable Sheffield-size keels to carry loads of about 90 tons, despite their limited length and beam.

The remarkable sight of a Humber keel under sail can again be experienced following the preservation of the keel Comrade and her re-rigging after many years as a motor vessel. Comrade was built of steel to Sheffield size in 1923, and for the first eleven years or so of her existence carried sail alone. With the passage of time an engine was fitted and, later, sails removed, and with successive engines she continued to trade until 1975.

By then she had been purchased by the Humber Keel & Sloop Preservation Society. This had been founded in 1970 with the aim of preserving

examples of vessels of both types, and sailing them in their home waters. By 1973 it was in touch with Captain Fred Schofield, owner of *Comrade*, who was happy to sell her to the society for preservation. In 1975 after carrying her last cargo she collected her new mast, delivered from Sweden to Gunness on the Trent, and took it to Beverley where society members, aided by Capt. Schofield and a job creation scheme, restored and rigged her. During the summer of 1978 she toured Yorkshire waterways under sail, her hold carrying an extensive exhibition of photographs of keels in days gone by.

Other keels of similar size have been preserved in public. That useful guide *Old Ships, Boats & Maritime Museums* records the 1898-built wooden hulled keel *Annie Maud* on exhibition on the Ouse at York, and the keel *Hegaro* was moored on the Thames at Greenwich, during 1978, with her hold converted into a childrens' bookshop. Her stubby shape made a marked contrast with the *Cutty Sark* nearby.

The keel *Ethel*, now an exhibit at the Boat Museum, Ellesmere Port, is a timber vessel built as recently as 1952 to carry coal on the Calder & Hebble Navigation. She was a motor barge from the start, and was probably the last wooden craft to carry commercially on inland waterways, for she remained at work until 1975 when she was bought by the Boat Museum with the help of a fifty per cent grant from the Science Museum. She reached Ellesmere Port from the Calder & Hebble by water, traversing the Leeds & Liverpool Canal on the way: I gather her passage of that canal's Foulridge summit tunnel was something of a cork-and-bottle situation!

No Severn trows have traded for many years, but in 1977 the Severn Trow Preservation Society was formed to preserve the trow *Spry*. This trow was built in 1894, originally as a sailing vessel, and later used as a dumb barge. Later still she became a hulk at Diglis Basin, Worcester, where the Worcester & Birmingham Canal joins the Severn. The intention of the society is to restore her to sailing condition or, failing that, as a static exhibit with an exhibition in her hold.

Unlike the Mersey, Humber and Severn, the upper Thames ceased to be used regularly by commercial traffic many years ago, and much of the traffic was in any case carried latterly by narrow boat. I do not know of any survivors of specifically Thames barges of the type which influenced the dimensions of connecting waterways such as the Thames & Severn. The nearest approach to this type of vessel which is preserved is probably the seventy-feet long barge *Aldershot*: this timber barge was built in 1931 at Ash Vale beside the Basingstoke Canal and was used for carrying until 1960. She was then sold and converted into a houseboat, and later purchased by the Surrey & Hampshire Canal Society for preservation.

Starvationers and narrow boats

The typical boat of the canals of the English Midlands was the narrow boat, and the forerunner of the narrow boat was the starvationer, the boat designed for and used in the Worsley coal-mine. Open boats, built of wood, with their ribs showing prominently (hence the nickname) they were little more than floating boxes with pointed ends. Sizes varied from about 30 feet long by 3 feet 4 inches beam up to 50 feet long by 7 feet beam. After mining ceased such boats continued to be used for tunnel maintenance at the mine and at least three survive—the one at Worsley mentioned in chapter ten, one belonging to the Boat Museum, and one at the Lound Hall Mining Museum, Retford, Nottinghamshire.

Boats of this type were used not only within the mine but also to carry coal from it to Manchester, and from them developed the narrow boat, approximately 70 feet long by 7 feet wide, to fit the locks of narrow canals. The earliest probably had only the simplest of cabins, or no cabins at all, and their appearance probably corresponded, among boats still extant today, to the day boats or 'Joey boats' used for short-haul traffic in the Birmingham area. Some of these remain in use—unpowered, they are hauled by horse or tug, and both the Black Country Museum and the Boat Museum have examples.

Those at the Black Country

Museum include a riveted-iron boat built in the 1890s by the Hartshill Iron Company, and a wooden version formerly owned by Stewarts & Lloyds and used to store and carry tubes. Pride of place, however, is taken by the restored Joey boat *Birchills*, built as recently as 1953 for carrier Ernest Thomas of Walsall, and restored for the museum, after seven or eight years of lying derelict, by M. E. Braine, boatbuilder of Norton Canes, who makes a speciality of this type of work. This boat has a small cabin fitted out for day use; she spent her working life carrying coal. She is double-ended—that is to say, the rudder can be mounted at either end of the boat, making her independent of winding holes.

The Boat Museum has two open day boats from the Birmingham area, one of them with an iron hull, the other a wooden Stewarts & Lloyd tube boat. Neither has a cabin.

Where narrow boats were to work over long distances, it became inevitable that they would be equipped with a sleeping cabin. At first, probably, they were worked by boatmen who had their homes ashore and slept aboard when away. Later, and particularly as an economy measure to meet railway competition, it became the practice for boatmen to live on board with their families, who helped to work the boats.

Thus evolved the traditional narrow boat, with long black hull and brightly painted upperworks, living cabin at the stern (and occasionally in the bows also, where a family had many children to accommodate), and most of the length of the boat taken up by the hold, with a capacity of about thirty tons. Over the hold are tarpaulin cloths; their ridge is formed by the top planks, running from cabin front to triangular upright *cratch* where hold meets foredeck. About one-third of the boat's length back from the bows, the towing mast protrudes through cloths and top plank; the stem of the boat carries a substantial rope work fender, the tapered stern an ornately decorated rudder.

8/7 (right) Rochdale Canal flats gently disintegrate in the canal in Manchester. The early railway-era viaduct overhead is perhaps symbolic.

8/8 Sheffield size keels at Waddington's yard on the Dearne & Dove Canal at Swinton Junction.

This general layout became almost invariable, though there were many differences of detail and proportion from canal to canal and from boat-builder to boatbuilder.

The living cabin of a narrow boat is about 10 feet long by 6 feet 6 inches wide. It is extremely cramped by modern standards, though possibly less so by the standards of early nineteenth-century slums. Boat people seem to have discovered by the middle of the last century—or probably earlier—how to furnish this interior with maximum economy of space, and to have stuck to this traditional layout ever since.

The sides of the cabin have much tumble-home, to prevent damage from bridge arches; the cabin is entered from the short stern deck through double doors which open outwards and fold against the cabin back. A roof hatch slides forward to allow people to step down into the cabin, where a step doubles as coal box—for immediately on the left of the entrance is the stove, in early boats of pot-bellied type, in later ones a more elaborate range. In cold weather the steerer can stand on the step, close the doors behind him or her, and keep legs warm by the stove while a well-muffled body projects through the hatch. From this position it is still possible to reach the tiller.

Forward of the stove is a food

cupboard with door hinged at its bottom edge: this enables it to open downwards to form a table. Opposite is a wooden bench seat which becomes a child's bed at night. Forward of these in turn is the main bed, across the middle of the cabin; the centre part of it folds out of the way in daytime to give access to a door leading to the hold. All spare space seems to be occupied by drawers.

From the cabin roof projects the stove chimney, which boatmen proudly decorated with three bands of brass; a chain restrains it if knocked from its seating by bridge or overhanging branch. Forward of it were kept the water cans, which contained the

supply of water for drinking and washing, with a mop resting with its handle through one of the cans.

In these circumstances generations of boat people lived their entire lives. The layout of a narrow-boat cabin, both interior and exterior, can be studied without fear of giving offence to occupants at the Waterways Museum, Stoke Bruerne, where an authentic replica is one of the most popular exhibits. There is more about it in chapter ten.

Here too can be studied at close range the traditional painted decoration of the narrow boat cabin and stern end. Narrow boat painting is bright, gay, in simple colours—greens,

reds, yellows—and for a very long time has always been done, according to a set of conventions, in largely stylised form: the details, nevertheless, gave scope for variation by individual painters. The most remarkable features are the landscape paintings: each includes a castle, a lake or river, a bridge and, usually, a range of mountains in the background; they occupy panels at the tops of the insides of the cabin doors, so that they can be seen from outside when the doors are folded open, and on the cupboard door that folds down to become a table. The castle landscapes were also often repeated—as on the museum example—on the cabin exterior, on panels at the stern ends of the cabin sides. Elsewhere are paintings of bunches of stylised flowers, mostly roses, which occupy (among many other locations) the lower part of the cabin doors, and the water cans.

The name of the boat appears on her 'top planks' at the stern, the name of her owner, or owning company, appears in large ornate letters on the cabin sides, and elsewhere about the bow and stern of the boat are numerous geometric or similar patterns— diamonds, circles, hearts and so on— picked out in bright colours.

The origin of this traditional painting is forgotten: it was certainly extant as early as 1858, for in that year John Hollingshead wrote an article *On the Canal* for Charles Dickens's magazine *Household Words* and in it described both decorations and cabin layout minutely: they were already common in the forms which remain familiar. Most probably the roses-and-castles paintings had their origins in commonplace Victorian decorative motifs, now forgotten, which were adapted to a specialised situation on boats and survived there because of the continuing preferences of the more-or-less closed community of boat people. The Boat Museum exhibits various early nineteenth century romantic landscapes and pieces of decorated enamelware with themes from which rose-and-castle decoration might well have originated. Hollingshead describes a canal-boat castle landscape as 'painted after the style of the great teaboard school of art', and while we cannot now tell precisely to what this piece of facetiousness refers, it was clearly some form of popular decorative fashion familiar enough to his readers to be gently mocked. As for the geometrical designs on narrow boats, they seem to me to have much in common with the decorations of old wooden bodied road and farm vehicles; similar types of pattern survive on the barrows of London costermongers and the carts of rag-and-bone men. The subject of narrow boat painting has had an entire and excellent book devoted to it (*Narrow Boat Painting* by A. J. Lewery) to which readers are referred for more information.

Steam-driven narrow boats were operating by the early 1860s, but never became popular, because the steam plant needed constant attention and took up so much space that it left too little room for cargo. They were mainly used on the Grand Junction Canal and canals radiating from it, for on the wide GJC they could conveniently tow an unpowered boat as butty and pass through locks together with it. The traditional narrow boat design was adapted to steam power by extending the cabin forward as an engine room, and building a rounded counter over the propeller, which emerged from the tapered stern, to prevent cavitation. Motor narrow boats, when they came into general use, followed the same pattern: but since the engine room was small and no engineer was needed, they were economical to run and adoption was quick and widespread.

The two Cadbury motor boats which pioneered use of the Bolinder engine seem to have represented a brave attempt to develop the design of boats for narrow canals beyond the point at which it had settled. Their hulls were built from mild steel plate, twin rudders were fitted to minimise wash and steering was by wheel instead of tiller. The holds were extended upwards above deck level and covered by a watertight steel deck containing hatchways for access to cargo. The cabins had side windows.

The engines gave satisfaction and the boats when new handled well: but in general these boats do not seem to have been satisfactory. The overall concept was sound, but there were deficiencies in detail. The builder's workmanship was bad and, I am inclined to think, being designed by naval architects these boats were not robust enough for canal conditions— constant passage of locks and contact with quays and other boats. Those cabin windows proved irresistible to boys with stones. Though *Bournville I* and *II* did useful work, they were not kept in service for very long, and were replaced by motor boats of traditional type with wooden hulls.

Because it was the general practice for motor narrow boats to tow butties, some boats originally intended for horse-haulage have survived, and their design was perpetuated in other unpowered boats built in the 1930s, 1940s and 1950s as butties for motor boats. One narrow boat that was certainly never regularly hauled during her trading life, however, by anything other than animal power is the *Friendship*, preserved at the Boat Museum. She was built of wood in 1925 for owner-boatman (or 'Number One') Joseph Skinner who used a mule for power and carried coal or timber on the Coventry and Oxford Canals. After retirement he and Mrs Skinner continued to live on board until their deaths in 1975 and 1976 respectively, after which, following an appeal for funds, the boat was purchased for the museum.

Two Fellows Morton & Clayton horse boats are preserved in public. One of these is *Vienna*, an iron composite boat built in 1911 and restored to her original appearance in the early 1970s. She is normally moored on the Caldon Canal at Cheddleton Flint Mill wharf, where she represents canal transport as formerly used by the mill, which is now a working industrial monument. The other is *Northwich*, which is an open-air, out-of-the-water exhibit at the Waterways Museum, Stoke Bruerne. She was built in 1898 and has a fore cabin as well as the usual stern cabin.

The last wooden narrow boat to be built was the *Raymond*, launched at Nurser's yard at Braunston in 1958. This boat continued in service until regular commercial traffic on the Grand Union expired in 1970. In 1978, her former captain Arthur Bray still lived aboard at Braunston, but no longer travelled.

Former horse-drawn narrow boat *Gifford*, preserved at the Boat Museum, was also built at Braunston, but earlier, in 1926. She was one of several specialised narrow boats with the holds decked over to carry tar or oil, forming part of the fleet of Thomas Clayton (Oldbury) Ltd. For part of her working life she was butty to a similar motor boat.

The best-known fleet of steam narrow boats was that of Fellows Morton & Clayton Ltd. Although FMC had no steamers in service after 1927, many of them had been converted to motor boats and some of these survive. One of them is *President*, built in 1909 and converted to diesel in 1925: she was sold by FMC in 1944. By 1973 she had been abandoned, sunk, in the River Weaver, but, however, was purchased

8/9 (right) Boatwoman using traditionally painted canal ware, showing how water could be poured from its usual position forward of the stove chimney.

by Malcolm Braine and Nicholas Bostock in partnership and taken to Braine's yard at Norton Canes on the BCN. There, over a period of four years, she was gradually and immaculately restored to her original appearance. Steam engine and boiler were re-installed, of types which are appropriate, though not identical to the originals, none of which survive. She was awarded the trophy for the best turned-out working boat at the IWA 1978 National Rally, and it is intended to exhibit her from time to time at museums on the canal system.

Sister FMC steam narrow boat *Viceroy*, built in 1909 and the last to be converted to diesel, has been converted back to steam; her hold now seats forty eight passengers and the boat is chartered out on the Kennet & Avon Canal at Bathampton by John Knill & Sons. Another FMC steamer, *Monarch* of 1908, has been restored to its original appearance externally— though it still has a diesel engine—and is exhibited at the Boat Museum.

I do not think any motor narrow boat has yet become a permanent exhibit in a museum: fortunately enough survive for their appearance to be familiar. They are used for maintenance, for the retail coal trade and occasional carriage of other freight, and for camping; many are owned privately, either converted or preserved in original state.

Commonest probably are survivors of the Grand Union Canal Carrying Co.'s fleet, which was built up during the thirties: in some cases they retain contemporary butties. These boats were mostly built of steel, though some were composites. They are bluff-bowed, deep hulled, maximum capacity boats intended to operate mainly over wide canals. The Narrow Boat Trust has two, *Alton* and *Nuneaton*, both all-steel town class boats built in 1936. The trust was established in the early 1970s to restore and preserve carrying narrow boats and to promote use of them for commercial carrying.

Among privately-owned ex-G.U.C.C.Co. boats, the pair *Towcester* and *Bude* owned by T. & D. Murrell are well known. At the time of writing

they are carrying grain on the River Wey; previously they have been used for retail coal selling on the Grand Union. Union Canal Carriers of Braunston have half a dozen pairs of narrow boats used for camping holidays, mainly by school parties and youth groups, and some similar single motor boats hired out on a self-steer basis. These include several G.U.C.C.Co. boats.

Fellows Morton & Clayton Ltd's boats had finer lines than G.U.C.C.Co. boats; several survive, including *Jaguar*, a pioneer of the retail coal traffic, and others used as camping boats. Among the last of narrow boats to be designed were the

Admiralty class, six pairs built in the early 1960s for the waterways of the North-West Midlands. They were ungainly vessels of steel, with steel hoops to support the top cloths and curious down-turned nose. I noticed one carrying piles for bank protection on the Shropshire Union Canal in 1978 and probably others are still in use.

There were several variants of the narrow boat, extra-narrow or extra-short, for instance, according to the canals to be traversed, but few examples survive. One that does is the 'Hampton boat, an outsize day boat formerly used exclusively on the Wolverhampton level of the BCN. There was sufficient internal traffic on this

8/10 Restored Fellows Morton & Clayton Ltd horse-drawn narrow boat Vienna *at Cheddleton Flint Mill, Caldon Canal.*

level to justify construction of these 87 feet long by 8 feet wide boats, which carried 50 tons of coal but were too wide to pass through locks. Fortunately the Black Country Museum is located on this level and has been able to preserve at least one of them.

The Bridgewater and Manchester, Bolton & Bury canals carried coal by boat in containers, a system which originated with Brindley. Some of the MB & B's container boats were still extant in 1978, sunk in the canal at grid reference SD 759066, where they could be discerned on the bed of the canal complete with rows of containers, provided there was not too much ripple on the surface.

Tub boats

From the starvationer developed not only the narrow boat but also the tub boat. Having now been disused for many years, they are much rarer than narrow boats, but at least three are preserved.

One of these is a wrought iron boat formerly used on the Shropshire tub-boat canals, which is now preserved on the Shropshire Canal within Blists Hill Open Air Museum at Ironbridge. It was found doing duty as a water tank on a farm, and re-floated in 1972.

While the Shropshire Canal tub boats were, quite literally, rectangular

tubs, those of the Bude Canal had the remarkable refinement of being fitted with wheels. On these they were raised and lowered up and down the canal's inclined planes without the need for cradles. One of these boats is preserved in the Exeter Maritime Museum beside the canal basin there. There is another Bude Canal tub boat preserved on the quayside at Bude; this lacks wheels but has one end pointed, and was used to lead a train of boats.

Wide boats and barges

The South Wales canals generally used boats 60 to 65 feet long and about 9 feet beam. A double-ended day boat of slightly smaller dimensions (58 feet by 8 feet) was raised from the bed of the Neath Canal early in 1977 and is now exhibited at the Welsh Industrial and Maritime Museum, Bute Street, Cardiff. It was built as recently as 1934, of wood, and used for maintenance until the early 1960s.

After the narrow boat, however, the specialised canal craft that were most widely used were the short boats of the Leeds & Liverpool Canal and the boats or barges of the Grand Canal. Both had similar dimensions—Leeds & Liverpool boats being 62 feet long by about 14 feet breadth, and Grand Canal boats 60 feet by 13 feet, and both, though they operated far apart, had similar origins. Leeds & Liverpool boats were derived from the flats of the Douglas and Mersey and the keels of the Yorkshire rivers; Grand Canal boat dimensions were based on those of the Aire & Calder at Smeaton's suggestion.

Leeds & Liverpool wide boats went through the same progressions from wood to steel and horse to motor as other craft, though steam wide boats were more popular than steam narrow boats and lasted into the 1950s. Some horse-drawn boats had transom sterns, some were rounded; cabins were provided in bow and stern, and some boats were crewed by families. Many L & L boats were highly decorated with elaborate scrollwork and floral decoration, though to less restrictive conventions than narrow boats.

Capacity of L & L boats is 40 or 50

8/11 (opposite) An 87 feet by 8 feet 'Hampton boat, next to an open day boat of normal size. Beyond is an icebreaker; left are limekilns. Black Country Museum.

8/12 (above) Tub boat and icebreaker on the Shropshire Canal within Blists Hill Open Air Museum, Ironbridge.

tons and there has been a mild revival of traffic on the Leeds & Liverpool during 1978. Wide boats may be seen carrying fish meal, grain, or sewage effluent. The Boat Museum has the dumb wide boat *George*, one of very few surviving transom-sterned examples. She (he?) is timber-built with a metal girder kelson, a typical feature of the type. She probably spent her life carrying coal. A similar but larger vessel at the Boat Museum is the L & L long boat *Scorpio*: she is 72 feet long, built for use west of Wigan on those sections of the L & L and adjoining canals which had long wide locks.

Grand Canal barges are generally so-called today, though the company and its boatmen referred to them as boats. Most survivors are all-steel motor craft of a type introduced in 1925, which largely superseded earlier

vessels. Those which survive are generally used for maintenance or converted to pleasure craft. They were worked by all-male crews; a living cabin was provided in the bows.

Puffers

Wide horse-drawn barges known as scows used the Forth & Clyde Canal, and so did sailing gabbarts not dissimilar to Mersey flats, but the typical trading vessel of the canal was the steam lighter or 'puffer'—so-called because early examples, instead of condensing used steam, exhausted it to atmosphere, locomotive-style, with appropriate noises. Such vessels, screw-driven, came into use in the late 1850s, and continued to be used until the canal closed, although a few were latterly converted from steam to diesel. They had developed into 'inside' puffers, without bulwarks, which were used only on the canal, and seagoing vessels which traded also on the West and East Coasts, and through the Crinan Canal, but were known from their principal habitat as Clyde puffers. The design was adopted during the Second World War for Admiralty victuallers and two of these, their holds converted into passenger accomodation, survive in the region where puffers were once common.

They are *VIC 32*, based as I write at Crinan Harbour on the canal of that name, and *Auld Reekie* (originally *VIC 27*), based on Oban. These surviving puffers still display the stumpiness of vessels less than 67 feet long, the length dictated by Forth & Clyde locks.

Passenger packet boats

Surviving relics of the actual boats are few indeed, though traces of the facilities they needed are less rare, and mentioned in the next chapter.

The last passenger packet boat in England was the Bridgewater Canal's *Duchess-Countess* which survived, as a houseboat latterly drawn up on the bank, until the late 1950s. Before being dismantled (or perhaps disintegrating) she was fortunately fully surveyed and photographed, and the resulting drawings and photographs are displayed in the Waterways Museum along with a few fragments such as her painted name board and the scythe-like knife, mounted on her bows, with which, as a packet boat, she was entitled to slice through the tow-rope of any lesser vessel which got in her way.

The hull of the Grand Canal fly boat (i.e., passenger packet boat) *Hibernia*,

8/13 (above) Steam tug
Mayflower *hauls barges along the Gloucester & Sharpness Canal, probably during the late 1940s.*

8/14 (opposite) Loading compartment boats at Castleford Aire & Calder Navigation. Each boat takes the load of two such lorries.

built in 1832, survived sunk in the canal until recent years, when it was raised. Its condition when I looked at it in 1978 was a disappointment: cut into three pieces, I believe, when raised, it had subsequently disintegrated into at least ten, stored beside the car park at Robertstown. Enough remained to show how light the construction was for an iron boat, with $\frac{1}{8}$ inch thick sheet iron plates supported by ribs of $1\frac{1}{2}$ inch by $1\frac{1}{2}$ inch angle iron typically 28 inches apart. It is to be hoped that this

apparent heap of rusty iron will be preserved better than when I saw it, for it is a unique survival of a unique era.

Tugs and maintenance craft

Steam and, later, motor tugs were used in two distinct ways on canals: to haul unpowered boats along long lock-free pounds, and to haul horse boats through tunnels without towpaths. In both cases they often hauled boats in trains, several at a time.

Canals where tugs were used in the first way included the Bridgewater, the BCN, the Ashby, the Grand Canal and the Caledonian. On the latter canal

they were in use in the early years to tow sailing ships when the wind was unsuitable, and they were used for the same purpose on the Gloucester & Sharpness, and also to haul trains of dumb barges along it. The 1861 steam tug *Mayflower* survived in poor condition in Gloucester Docks until recently, when it was offered for sale by auction: I have yet to hear the outcome. Some BCN motor tugs are still in use to haul day boats, and others are privately preserved.

In 1908 H. R. de Salis was able to say that 'in modern times' where there was any volume of traffic through a tunnel without a towpath, steam tugs generally towed boats through in trains at fixed times. Such services did not long outlive the general introduction of self-propelled motor boats. A rare survivor of that era is the Boat Museum's tug *Worcester*, built in 1912 for the Worcester & Birmingham

Canal, which used her to tow boats through the long tunnels above Tardebigge and along the intervening open-air sections. She has always been a motor vessel—her present engine is a 1928 30 bhp Bolinder semi-diesel—but her design represents a transitional stage between steam and motor. She has a capacious engine room and exhausts up a funnel. Her builders, Abdela & Mitchell of Brimscombe on the Thames & Severn, were builders of steam launches and tugs and gave her an elegant hull with a counter stern.

From haulage of keels on the Aire & Calder was developed that waterway's compartment-boat system, as mentioned in chapter six. It remains in use, though trains are now towed by motor tugs rather than propelled by steam ones. Capacity of boats has been increased to about thirty five tons and they are towed in trains of about twenty. This century-old system is still

competitive against more modern forms of transport, such is the inherent economy of water transport for bulk commodities. Loading points where compartment boats may be seen are Doncaster (S & SYN) and Castleford (A & C).

Many canals had elegant steam-driven inspection launches. One of the few which survives is *Sabrina*, originally used on the Gloucester & Sharpness Canal. She later became a houseboat on the Thames, but has now been restored to steam power by her present owners. Steam power was also used for dredgers, as early as 1816 during construction of the Caledonian Canal. A steam dredger which survives in use is the Surrey & Hampshire Canal Society's floating grab dredger, originally on the Grand Union, which is now assisting with restoration of the Basingstoke Canal.

Vital to canals in the past were ice-

breakers, whether horse-drawn, steam or motor. Being sturdily built and seeing only intermittent use, they are long-lived and correspondents D. N. and J. P. Wells, in a letter to the Editor of Waterways World (May 1978) stated that they knew of at least twenty iron and wooden ice-breakers still afloat. The Boat Museum, the Blists Hill Open Air Museum and the Black Country Museum all have them. The latter has a good example of a horse-drawn ice-breaker, dating from 1868 and formerly used on the BCN. She is built of oak with her hull reinforced externally by iron plating; two short masts formerly supported a lengthwise beam to which gangs of men could hold while rocking her to break ice, and the hull itself has a supremely elegant pointed shape born of the practical necessity that it should ride up on to the ice and then smash down through it.

8/15 Sabrina *was originally the steam inspection launch of the Gloucester & Sharpness Canal. After many years as a houseboat, she was re-equipped with steam plant by her present owners, and is seen here on the Thames at Oxford during a cruise organised by the Steam Boat Association of Great Britain.*

CHAPTER 9

FEATURES OF CANALS: THE SURROUNDINGS

Wharves, cranes and warehouses

If today's pleasure boaters ever tend to forget canals' commercial past, they must surely be quickly reminded of it by the frequency with which former commercial wharves appear along the canalside during the course of a cruise. They are now commoner in the country than in towns, for in towns they have presumably more redevelopment value. Typically, the village wharf (which might be some distance from the village it served) comprised a walled-in area, where coal or timber might be stored, on the side of the canal opposite the towpath, with a crane and a warehouse.

In some places they still give the illusion of being in use, for coal merchants are slow to move away and continue to use canalside wharves even though receiving their coal by road. Examples that come to mind are the wharves at Cromford, Cropredy (Oxford Canal) and Silsden (Leeds & Liverpool). There are many others. Occasionally wharves, briefly, come back to commercial life. On a visit to Gayton wharf (Grand Union) I found the pleasure boats that usually occupy it moved aside to make room for retail coal boats *Towcester* and *Bude*. Other former wharves continue to be used for commercial purposes unconnected with the canal—Stretton Wharf, for instance, on one of the Oxford Canal loops, has still a transport significance: it is used as a lorry park. The happiest form of re-use is surely found at those wharves which have become bases for hire fleets—just two examples are

Aynho (Oxford Canal) and Wootton Wawen (Stratford).

Canal warehouses were generally built of brick or stone according to region, and have a pleasing appearance, plain but well-proportioned. Usually they are close beside the canal, sometimes an awning projects over it to shelter boats being unloaded, occasionally an arm of the canal leads inside the building.

In large towns, canal basins were built: spurs off the main canal (or perhaps its termination), wider than usual, and lined with wharves and warehouses. In London, City Road and Paddington Basins were the main terminals for canal traffic: both are still in water. At the far end of the canal map, Ripon basin survives, unnavigable and a nesting place for swans. The extensive basins at Stourport are now used for pleasure craft moorings, that at Nantwich is the base for one of BWB's hire cruiser fleets. There were many others. In Ireland, such canal basins were known as harbours—the large harvour at Tullamore survives, though it has lost some of its buildings, while James's

9/1 Many coal merchants still use canal wharves, even though their coal is carried by road. This wharf is at Silsden on the Leeds & Liverpool Canal.

Street Harbour, the canal terminal in Dublin, has recently been filled in, although many warehouses, offices, houses and other canal buildings still stand, or did in 1978. In Scotland, canal basins were still further dignified by the title 'Port'. The observant in Lothian Road, Edinburgh, may note a barge carved in relief on the facade of a 1930s block of shops and offices, to indicate that it occupies the site of Port Hopetoun, former terminus of the Union Canal.

At some basins canals connected with feeder tramroads. One of the most extensive was at Buxworth, where the Peak Forest Canal met its tramroad extension. These was a complex of basins, wharves and warehouses more than half-a-mile long which became disused after about 1926. Since 1968 the Inland Waterways Protection Society has been steadily reclaiming them from overgrown dereliction and restoring them as a labour of love, work which is, at the time of writing, most regrettably threatened by proposals to construct a bypass road across the site.

One hopes that the outcome of these proposals will be as happy, from the canal point-of-view, as those of the early 1970s for demolition of the warehouse at nearby Whaley Bridge (grid reference SK 012816). This warehouse was laid out for transfer of goods between the canal and the Cromford & High Peak Railway. The canal entered it at one end, and within were railway tracks along each side of the canal, emerging from the far end. I mentioned proposals for its demolition in *Waterways Restored* (1974), and so am happy to be able to report here that it has survived and is now the base of boatbuilder Coles Morton Marine, and includes a small waterway museum.

High Peak Wharf on the Cromford Canal was the canal/rail interchange point at the other end of the C & HPR; when I was last there, some years ago, it retained several early buildings; and Froghall basin, terminus of the Caldon Canal, was the interchange point for the tramroad from Caldon Low quarries.

To find actual tramroad relics in position beside a canal is rare. Blists Hill Open Air Museum, Ironbridge, has reinstated a section of tramroad alongside its canal, but the most interesting relic of this type is undoubtedly the primitive wagon of the Stratford & Moreton Railway which stands, preserved, on a short length of original track beside the canal basin at Stratford-upon-Avon. Of the later era of railway interchange basins on the BCN, several examples remain in water; there is a good one at Great Bridge (grid reference SO 978927).

Canal villages and towns

Canal villages, where the canal forms a linear centre-piece and many of the buildings, connected with it, line its banks, are one of the most attractive features of canals—and also, so low was the esteem in which canals were generally held throughout much of their life, one of the rarest. Some of them were villages long before canals came, and adjusted themselves to suit what must have been an overwhelming newcomer. Stoke Bruerne, on the Grand Junction Canal, is the prime example of this type of village, and is described in chapter ten. Robertstown on the Grand Canal is its equally attractive Irish equivalent, and at Fort Augustus on the Caledonian Canal shops and houses line each side of the gigantic staircase flight of locks.

More interesting still are those settlements which sprang up, as a consequence of construction of canals, independently of any previously-existing village. Usually they were built where one waterway met another. Fradley Junction (grid reference SK 141140) is a mile and a half from Fradley village, but is one of the nodal points of the canal system, where the Coventry Canal (forming the south eastern leg of Brindley's Cross) meets the Trent & Mersey. So all its buildings are canal oriented: opposite the actual junction an eighteenth-century row includes cottages, pub and warehouses

(which are now the hire base of Swan Line Cruisers Ltd); in the angle between the Coventry Canal and the eastern part of the T & M is Georgian *Wharf House*, once occupied by the Coventry Canal manager, with a room on the ground floor reserved as toll keeper's office; in the other angle between the two canals is a dry dock, original although the building over it is new and in keeping; in the vicinity are lock cottages and a maintenance depot. Access is by tracks alongside the canal, the towpath being upgraded briefly into a roughish road.

At Shardlow, to the east along the Trent & Mersey Canal and a mile short of the point at which it enters the Trent, a canal village grew up near to but distinct from the original one. Here, goods were transferred between Trent barge and narrow boat. Basins, pubs, wharves and warehouses remain along the canal, and the long filled-in arm which once took boats inside a handsome corn mill for unloading has been re-excavated so that the mill may become a hire base.

Shannon Harbour served the same purpose: here, goods were transhipped between Grand Canal boats and more seaworthy vessels better able to navigate the Shannon lakes. It is, almost entirely, a canal settlement: the canal widens and along its south bank are ranged transhipment sheds, stores, the agent's house, ruins of a hotel for canal passengers, and other canal buildings.

Stourport, built where the Staffordshire & Worcestershire Canal met the Severn and goods had to be transhipped between narrow boat and trow, has grown into a town. Its centre remains the complex of basins in which the canal terminated—though this is less complex than formerly since the most easterly basins were filled in after commercial traffic ceased but before the pleasure craft boom started. Around, among and between the basins are many canal buildings in mellow red brick, notable among them being the central warehouse surmounted by clock tower and weather vane, the latter doubtless provided for the benefit of skippers of sailing trows.

9/2 Irish canal village: Robertstown, beside the Grand Canal. In the foreground, a former Grand Canal Co. motor barge is used for maintenance.

9/3 (above) Canal town: Stourport on 12 April 1926. The third basin, to the right of the others, can be seen, and at its right hand edge the coal wharf for the power station. This basin was filled in after commercial traffic ceased but before the pleasure craft boom began – boats of all sorts are few in the picture. Several canalside buildings, since demolished, can also be seen.

9/4 (right) Stourport on 13 July 1970. The Staffordshire & Worcestershire Canal approaches from the top right hand corner and enters the basins; duplicated wide and narrow locks (intended for trows and narrow boats respectively) lead down from the basins to the River Severn in the foreground. Parked cars and industrial buildings occupy the site of the third basin, now filled in, to the right.

The Tontine Hotel, between basins and river, was built by the canal company which held its meetings there; it is often said to have been used by canal passengers (although I have been unable to find out precisely what services were operated), and it was certainly used by business visitors to the town. It recently escaped threatened demolition, and, when I saw it in 1978, it was surrounded by scaffolding for repairs.

Equally fascinating is Ellesmere Port. Its period of greatest development came later than most such places, only after the Birmingham & Liverpool Junction Canal was opened in 1835. Then, because the Shropshire Union company maintained a degree of independence from its overlord the London & North Western Railway, and because as a matter of policy goods were transferred between narrow boat and Mersey flat at Ellesmere Port rather than further inland, a great transhipment port grew up there.

Its basins are at three levels, the highest being that of the canal from Chester and the lowest, once fully tidal, is now that of the Manchester Ship Canal to which it has direct access. Beside the entrance from the ship canal, a small brick lighthouse stands, in memory of the days when this was the entrance from the Mersey estuary. Two pairs of parallel wide and narrow locks lead from top to middle level, a barge lock and a ship lock connect the middle level with the lowest.

The former toll office, a large building for its purpose which stands near the top set of locks, has been restored by the Boat Museum, a voluntary body, as its headquarters. A large grain warehouse on an island at the top level is being restored with the intention that it should house the undercover exhibits of the museum. It includes a steam-driven hydraulic pump house built in 1876 which is being restored complete with boilers and pumps.

Cranes and other equipment at these docks were once operated by hydraulic power, that favourite Victorian form of mains power which is now, since the advent of electricity, almost forgotten.

The museum boats, described according to their various types in chapter eight, are kept in the basin on this level. From a vantage point beside it much of the rest of the docks and their buildings—warehouses and workshops—can be seen. Many of the buildings are derelict but none the less fascinating—a range of boat-horse stables, for instance, complete with stalls and mangers. Most of the surviving buildings were listed in 1978 as being of special architectural or historic interest. This was too late, unfortunately, for the most important warehouses—these were destroyed some years ago by fire which is generally considered to have been arson. They had been designed by Telford, with access from both the top level and the middle level, where flats could float beneath them for loading and unloading; their foundations can still be seen in the form of a series of promontories in the middle basin. Housing in many streets nearby was built by the canal company, for many years the largest employer in the district.

Ellesmere Port on the west coast was matched by Goole on the east where the Aire & Calder Navigation built extensive docks for interchange of cargoes between barge and ship (and also, following agreement with the Lancashire & Yorkshire Railway, rail wagon and ship). It too built housing, and hydraulic hoists to raise compart-ment boats and tip their contents into waiting ship; and the port and its canal remain busy with commercial traffic.

The opposite, sadly, is the case with Grangemouth. This port grew up around the sea lock at the eastern end of the Forth & Clyde and owes its existence to the canal; in recent years it has turned its back on it and though the docks continue in use, the canal approach to them is filled in. Runcorn is similar. It was a hamlet until the Duke of Bridgewater built his canal to enter the Mersey there, after which it became a thriving interchange port. The docks, now on the Manchester Ship Canal, remain, but they are served by road. Both the flights of locks which formerly linked them with the Bridgewater Canal have been filled in.

Canal cottages and houses

Lock cottages, inhabited by canal employees, varied more than any other canal structure according to vernacular styles of architecture, from Cotswold stone houses on the Oxford Canal to typically Scottish cottages on the Forth & Clyde. Thomas Omer, however, embellished his early attempts to build the Grand Canal with some classical structures which survive at 11th and 12th Locks, and later lock cottages sometimes had Regency features—at the tops of the flights at Grindley Brook on the Ellesmere Canal, and on the Grand Junction's Northampton branch, for instance.

The Regent's Canal equipped Camden Lock with a curious castellated edifice. The lock cottage at Tyrley, Birmingham & Liverpool Junction, starts to show Gothic features, and that at the top of Perry Bar flight on the Tame Valley Canal is stern Victorian.

Of purely canal design are the round-tower-like lock cottages which the Thames & Severn built, and also the Staffs & Worcs at Gailey—they may have been built like this to improve outlook for approaching boats. Unique, too, are the strange barrel-vault roofed lock cottages of the southern section of the Stratford Canal. Their roofs consist of a length of brick arch similar to a bridge arch, and they are said to have been made by canal builders who were more familiar with this form of construction than with normal roofs. Several examples survive, notably at grid reference SP 177657.

Other small but unique and characterful buildings survive beside several locks of the Chester Canal: very small round buildings, brick built, with domed roofs and a door which continues the curve of the walls, they were intended mundanely for the use of lengthsmen maintaining the canal. There is a good one at Beeston Stone Lock (grid reference SJ 557596).

Most opening bridges are opened by boatmen, but this was not practicable on a ship canal such as the Gloucester & Berkeley. The little houses that were provided for the bridge keepers are perhaps the most consciously elegant series of buildings related to canals. They are small houses in the classical

9/5 (opposite) The interior of canal boat-horse stables, complete with stalls, mangers and cobbled floor, derelict at Ellesmere Port in 1978. Patterns of sunshine and shade are caused by the holes where slates are missing from the roof.
9/6 (above) The exterior of derelict canal boat-horse stables at Ellesmere Port, 1978. The building is said to date from 1801.

Warehouses of British Waterways Board's Broad Street Depot beside the main line of the Birmingham Canal Navigations at Wolverhampton. An arm of the canal, now closed by a gate, led boats into the warehouses for loading and unloading. The cast-iron roving bridge which carries the towpath over the canal is typical of many built on the BCN between 1820 and 1850. Easily-graded ramps led horses up onto it, and its parapets originate at ground level so as to present no obstruction to tow-lines.

Wharf cranes were once commonplace beside canals. Today they are generally found to be either derelict, with many of their components missing, or else brightly painted as canalside ornaments, which are equally unworkable. This one at Ellesmere Port is a rare example in workaday condition and still in use—though its task is now to unload scrap timber onto a bonfire.

Canalside industry: Coalport China Works. The works were established in the mid-1790s on a site between the then-new Shropshire Canal and the River Severn, in those days a busy navigation at this point. They provide a good example of how water communications attracted industry. The canal, a tub-boat canal which descended into Coalport by The Hay inclined plane, eventually became disused in the 1890s and production of Coalport china was moved to the Potteries in 1926; but both canal and china works have recently been restored by the Ironbridge Gorge Museum Trust.

Twin-arched bridges on Telford's new main line of the Birmingham Canal—this one is near Farmer's Bridge, Birmingham—were laid out to enable traffic in either direction, both boats and horses, to pass without hindrance. Each arch spans ample width of water for a narrow boat, and towpath as well. Horses need neither swerve nor duck, as they frequently had to to pass beneath earlier bridges. The canal itself is laid out on an even, flowing curve, in contrast to the tortuous windings of the earlier contour canal. Grid reference SP 057867.

9/7 (above) The Chester Canal provided ornate little circular brick-built huts for its maintenance men. This one is beside Beeston Stone Lock.

9/8 (right) Lock cottage with barrel-vault roof beside the Stratford Canal at Lowsonford. Cottages of this type are both unique to the southern Stratford Canal, and typical of it.

style, white-painted, with pediments at the front supported by two pillars. They date, probably, from the 1840s rather than the canal's impoverished early years.

To calculate tolls, boats were gauged: their freeboard was measured and compared with records held at toll offices of each boat's freeboard for given weights of cargo. The operation was conveniently carried out while the boat was in a lock, and was often done in stop locks where one canal joined another. So these and other locks are in many places accompanied by toll offices, which sometimes form part of a cottage or house and are sometimes independent buildings. There is a pleasant example of the former type— a white-painted house with a bow front so that those inside can see up and down the canal—at Pontymoile, where the Brecon & Abergavenny Canal joined the Monmouthshire. The canal still passes through the chamber of the stop lock in front of it.

Of toll offices which are independent buildings, The Bratch locks on the Staffordshire & Worcestershire Canal are surveyed by one which is a neat octagonal tower; at the extremities of the same canal, there is a small rectangular brick toll office with iron window frames in the Gothic manner at Great Haywood, and another rectangular office with bow window facing Stourport lock, above the basins, which dates from the 1850s and has become a canal-ware shop.

Where canals were exceptionally busy, toll offices were placed on islands in mid-canal. There is one between the paired locks at Kidsgrove on the Trent & Mersey, at the head of the long descent from the summit pound to the Cheshire Plain. The Birmingham Canal Navigations had octagonal toll houses on islands at intervals along its main line, the narrows either side being about seven feet wide and a boat's length long. The offices have all been demolished, but the islands remain to mystify visitors as to the purpose. Fortunately one toll house was care-

fully dismantled by the Birmingham Canal Navigations Society and it is intended to re-erect it at the Black Country Museum.

The Coventry Canal manager's house at Fradley Junction, which incorporated the toll office, has already been mentioned. Other canals had comparable buildings: the Union Canal provided an elegant residence, *Canal House*, at Linlithgow for its engineer, and the Ellesmere Canal built a handsome brick Georgian house as office at Ellesmere, opposite the entrance to the short branch leading to the basin; an extension at one corner, of almost circular plan, would have enabled those within to see down the canal in all three directions. The most impressive of such buildings, however, is Cleveland House, Bath (grid reference ST 758653). This was built of stone with all the architectural refinement to be expected in that city. It contained the offices of the Kennet & Avon canal, which passes, through a short tunnel, directly beneath it.

Pubs, hotels and traces of passenger traffic

From some market research which I carried out among hirers of canal cruisers during 1977, it was clear that some liked canal pubs and others loathed them. I cannot help wondering whether the latter's opinion sprang from convinced abstinence or simple disappointment. Originally the canal-side public house was the working boatman's place of recreation, and while few have survived unmodified internally, many remain in situ and active, serving the new generation of pleasure boaters.

Some of these pubs are even older than canals, and adapted themselves, when canals were built, to the then new form of transport. These include *The Boat* at Stoke Bruerne and the *Admiral Nelson* at Braunston, both beside the Grand Junction Canal, and (probably) the *Dog and Doublet* at Bodymoor Heath on the Birmingham & Fazeley. Others, from their locations among canal oriented buildings, were clearly built specifically to serve the canal—such are the *Swan* at Fradley Junction and the *Greyhound* at Hawkesbury. There are many more, of both types.

In some places, presence of a pub bearing a name with canal associations indicates the presence of a canal long closed or disused. There is a *Canal Tavern*, for instance, beside the basin of the Thames & Medway Canal at Gravesend, and a *Navigation Inn* near the basin of the Ripon Canal at Ripon. The *Union Inn* near Lock 16, Forth & Clyde Canal, once catered for canal passengers; close to it the Union Canal joined the F & C.

With the possible exception of the Tontine Hotel at Stourport, there was no equivalent in Britain of the hotels for canal passengers erected by canal companies in Ireland. Many of these, particularly those erected by the Grand Canal company, were fine buildings in the classical style; although they ceased to function as hotels about the time that passenger boats stopped running, several survive, the most important being those at Shannon Harbour, Robertstown and Portobello, Dublin.

The hotel building at Shannon Harbour is, as mentioned above, now derelict, though the structure is not so derelict as it at first appears, since it was deliberately made uninhabitable to avoid liability for rates.

Robertstown canal hotel might well have gone the same way: but through local enterprise it has survived. Every summer it forms the centre-piece of Robertstown canal festival; Georgian banquets are held there, and the upper floors are a museum. Many of the exhibits are to do with canals—evocative contemporary maps of might-

The original line of the Oxford Canal near Stretton Wharf is now a leafy backwater (grid reference SP 441809). Bypassed when the improved and shortened line was opened in 1834, this part of the earlier line was left in water as a short branch to serve the wharf. This is now disused and most of the branch is unnavigable, although near its junction with the later line it is used as moorings for pleasure craft which can be seen in the distance.

have-been canal extensions and a fine model of a Grand Canal motor barge being among them. The rooms of the top floor house a collection of full-size horse-drawn carriages (yes, the *top* floor), suggestive of a ship-in-bottle conundrum of how they got there. The Robertstown hotel building received an award during European Architectural Heritage Year 1975.

The building of the Portobello hotel in Dublin stands beside the one-time starting point of the passage boats, but for most of this century was a nursing home. Then, in 1971, it changed hands, and has since been renovated. Internally it is now used as offices, but externally it has been pleasingly and appropriately redecorated in grey picked out with white, and a cupola which old pictures show to have once embellished the roof has been reinstated.

These canal hotel buildings are by far the most substantial relics surviving from the era of canal passenger boats. But there are others. At Worsley is the Packet House, a half-timbered building probably older than the canal, where passengers purchased their tickets for the Duke of Bridgewater's packet boats to Manchester. Alongside it is the flight of stone steps which they descended to go on board. Not far away as the crow flies, at Nob End (grid reference SD 752064) can still be seen the cobbled quay to which the Manchester Bolton & Bury Canal's packet boat from Bolton moored. Here passengers walked down the pathway beside the adjacent locks, saving time and water, to join the Manchester boat which waited below, and no doubt passed en route other passengers from Manchester bound for Bolton.

Another place where passengers were encouraged to walk for a similar reason was at Falkirk, where the Union Canal fell by its sole flight of eleven locks to join the Forth & Clyde. To shorten the walking distance, a branch of the Union Canal was built, extending for 570 yards along the contour beyond the head of the locks to a new terminus called Port Maxwell. Today, the locks have been obliterated, and I have been unable to find the precise location of the junction with the branch: the canal appears continuous and, fringed with reeds, terminates abruptly at the roadside. There is nothing to indicate its purpose, nor to show that the final

quarter-mile of the canal is unique, a canal built for passengers rather than freight.

The Union Canal shows other traces of the passenger boat era, such as stable buildings which stand beside it at Woodcockdale (grid reference NS 977759). This was one of the stages at which the horses of passenger boats were changed.

Considering how common they must have been, it is surprising how rare canalside stables now are. I have already mentioned those at Ellesmere Port; there is another fine range beside the Shropshire Union Canal alongside the top of Bunbury staircase, although internally it has been adapted for boat building.

Boat-horse stables remain in use at Llangollen wharf, housing the horses which haul passenger boats, and an old-style stable was to be built in the Black Country Museum adjacent to a typical Black Country boatbuilder's yard which has been reconstructed there. It is in due course to contain an occupant.

Boatbuilding and maintenance yards

The Black Country Museum's boat dock or boatbuilder's yard may be a reconstruction, but it certainly has an authentic appearance and is typical of a great many such docks which once built and maintained the vast fleets of boats on the BCN, and which are now as rare as canalside stables.

Its period is the turn of the century, before motors started to oust the horse: a slipway adjoins the museum's canal arm, and on it boats can be drawn sideways out of the water. Behind it a range of simple buildings has been constructed. One of these is a blacksmith's shop, built to an 1880s design re-using old bricks. As soon as its bellows had been set up it was to become a working shop. Adjoining are several buildings made, typically, from the timbers of old boats—a nail store, a rivet shop, a paint store. A sawmill and a store for seasoning timber were

being added, as was the stable.

A barge yard which is still in use is Waddington's yard at Swinton, which, pending enlargement of the SSYN, itself appears to have been little altered. Here Sheffield size keels are maintained; the yard adjoins a short surviving length of the Dearne & Dove Canal, within sight of the SSYN at Swinton Junction. Some former commercial boatyards have outlived the end of commercial boating and build and/or maintain pleasure craft. Such are Taylor's boatyard at Chester on the Shropshire Union, and Tooley's yard on the Oxford Canal at Banbury which, when I last saw it, almost alone of the canal installations there had survived the ravages of municipal redevelopment. According to D. and J. Hadley, writing in *Waterways World* (December 1978), there has been a boatyard here for 190 years.

All canals had maintenance yards, of various degrees of elaboration, set at strategic points on their systems. Here men and materials for maintenance of the canal and its works were based. Unlike many canalside installations, they continue in many instances to serve their original purpose: Bulbourne and Gayton on the Grand Union, Tardebigge on the Worcester & Birmingham, Ellesmere and Norbury Junction on the Shropshire Union are examples. At such places lock gates and canal ironwork are made, and the typical maintenance yard comprises a pleasant and workmanlike jumble of buildings—some of canal era and some later—containing workshops for carpenters and blacksmiths, offices and stores, set around a yard stacked with timber and piles of bricks. There is a canal frontage lined with cranes, and to it are moored maintenance craft—dredgers, mud hopper boats, weed cutters.

Some yards had real architectural pretensions—Hartshill on the Coventry Canal, for instance. Here, between successive bridges, the passing boater catches a glimpse of canal arms which disappear into workaday Georgian buildings, one of which is surmounted by a handsome clock tower. Sometimes curious survivals are to be found. At Ellesmere maintenance yard there was in the spring of 1977 a handcart which was a clear reminder of the canal's former railway ownership: it was still lettered *LMS S.U. Canal Sct.* in faded gold over

9/9 (top left) Gauging a narrow boat on the Grand Union Canal to establish its displacement, and so the weight of cargo, from which the toll payable could be calculated. The instrument comprises a tube containing a float connected to the scale, from which the freeboard of the boat is read directly; a register of the displacement figures for various weights of cargo was held for each boat at each toll office on the canals where it worked.

9/10 (top right) A toll office survives beside the Staffordshire & Worcestershire Canal at Great Haywood.

9/11 (above) The Ellesmere Canal Company's offices beside the canal at Ellesmere.

The dry bed of the original course of the Oxford Canal at Newbold-on-Avon leads to the tunnel mouth in the distance. The tunnel passed beneath the churchyard, though why it was made is not clear, for it was nowhere far beneath the surface. Part of it remains in existence and part has been filled in. No trace can be found of the other portal. This part of the original canal was superseded by the improved line built in the early 1830s. Grid reference SP 487770.

The Grand Canal Company was noted for its passenger boats, or passage boats as they were called, and to accommodate passengers built several canalside hotels. These buildings ceased to be used as hotels after passenger traffic finished in the early 1850s, but several of them survive. This one is at Portobello, the Dublin terminus of the passage boats, which arrived at and departed from the quay in front of it. It is now used as offices but has been handsomely restored externally to its original design as far as is known. An earlier representation of the same view appears on page 53.

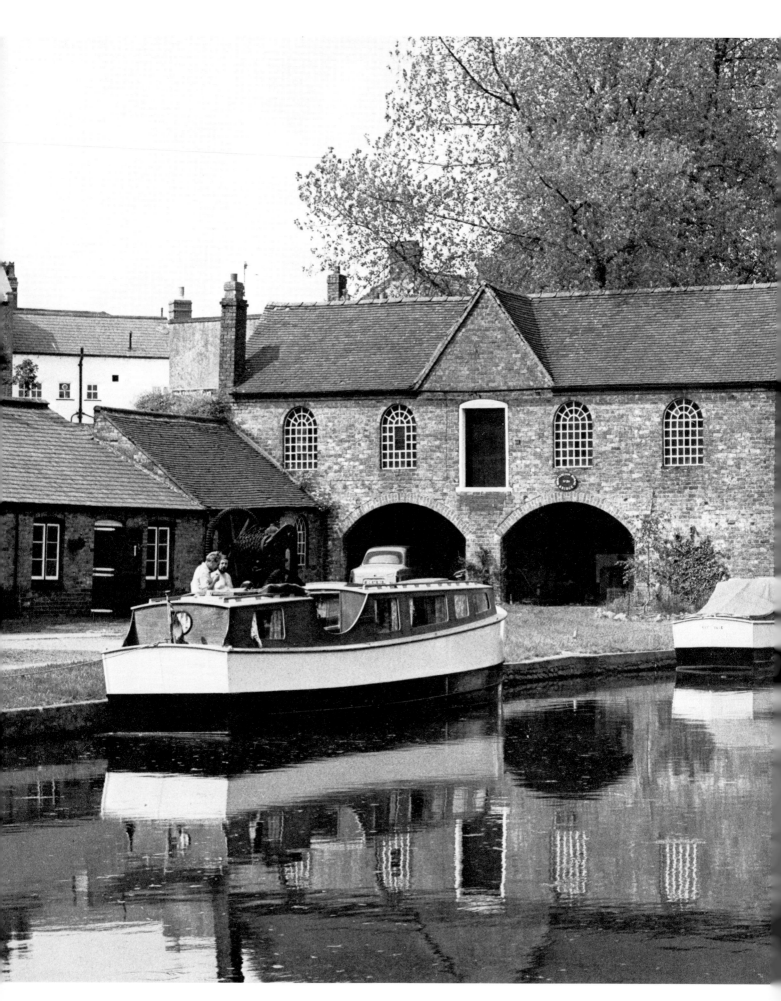

weathered maroon paint—the shades which, as railway minded readers know, were used by that pre-nationalisation company for its passenger coaches. For this vehicle still to display the paint of its former owner twenty nine years after nationalisation must have made it almost if not wholly unique.

Some maintenance yards, no longer needed as such, have been adapted to other uses. One of BWB's hire fleets is based at Hillmorton, at the former maintenance yard of the Oxford Canal; and a former Staffs & Worcs Canal maintenance yard at Stourport, which has particularly attractive buildings, has for long been used as the hire base of Canal Pleasurecraft Ltd, one of the pioneer hire operators.

Accompanying maintenance yards, or forming part of them, are often dry docks for maintenance of boats and barges. Sometimes they are covered in, sometimes open to the air; it was common practice to site them alongside locks so that water could be drained from them to the pound below. Occasionally they outlive canals: at Winwick Quay on the St Helens Canal (grid reference SJ 596917) it seemed in 1978 that the filling-in contract extended only to the main canal, for though that had been filled in and levelled, the dry dock leading off it remained—though that too seemed to be used as an unofficial rubbish tip.

Water supplies

Water in a canal is taken for granted until it runs short. This was brought home to me graphically when analysing the results of that 1977 market research survey among boat hirers which I mentioned earlier. One of the questions, framed during the 1976 drought year which, to save water, had many restrictions on the use of locks, was 'Was your cruise affected by water shortage?'. The answers, in many cases, were written in terms of the infrequency of canalside drinking water taps!

9/12 A former maintenance yard of the Staffordshire & Worcestershire Canal at Stourport is now used as a hire cruiser base.

British Waterways Board has about ninety reservoirs which supply water to its canals. Often they are remote from the canals they supply and their purpose is not obvious. Killington Reservoir in the fells of Cumbria, for instance, is overlooked by a service area of the M6 motorway: how many of those who pause to enjoy the view know that its function is to supply water to the Lancaster Canal? Similarly the purpose of Rudyard Reservoir in Staffordshire and the Welsh Harp (or, more properly, Brent Reservoir) in North West London is to provide water for canals, the Trent & Mersey and the Grand Union re-spectively. In each of these cases long feeder channels connect reservoir with canal, that from Rudyard being in part navigable as the Leek branch of the Trent & Mersey.

The Grand Union has another series of reservoirs on the northern slopes of the Chiltern Hills, some of them adjacent to the canal at Marsworth, some further afield. A complex system of headings and pumps enables water from them to be fed into the remaining section of the Wendover Arm of the canal, which connects with the summit pound of its main line. Further headings enable two of the reservoirs adjacent to the main canal to feed it directly or—in one case—to be fed with water coming down the canal, according to relative levels.

Elsewhere, water supply was occasionally obtained by leading a canal into a river and then out of it again—for example, the Trent & Mersey Canal and the River Trent below Alrewas, Staffs, and the Oxford Canal and the River Cherwell at Aynho and Shipton, Oxfordshire. The Ellesmere Canal obtained its water from the River Dee at Llantysilio above Llangollen, the Llangollen branch of the canal being in effect a partially-navigable feeder, of smaller dimensions than an ordinary narrow canal. In Ireland the Grand Canal's principal source is called Seven Springs, a pool from which water is led to the canal west of Robertstown by the eight-mile Milltown Feeder or Grand Supply. Like the Llangollen branch, this was made partially navigable.

9/13 A converted narrow boat in dry dock beside the Grand Union Canal at Hatton. In the left background can be seen the remains of a narrow lock, and to its left – beyond the electricity post – the wide lock which replaced it in the 1930s.

9/14 (above) Startopsend Reservoir, right, is one of the complex of reservoirs near Tring which feeds the Grand Union Canal, left.

9/15 (overleaf) The pump house which for many years contained a Newcomen engine is still prominent at Hawkesbury Junction beside the Coventry Canal, left. The Oxford Canal passes through the stop lock, right, and joins the Coventry Canal beneath the roving bridge, centre.

Where water could not be led into a canal by gravity, it had to be pumped in. Today this means electric pumps, but in the canal era it meant steam beam engines. The engine houses and, in a few instances, complete engines, can still be seen beside canals.

The Coventry Canal installed a Newcomen or atmospheric steam engine at Hawkesbury in 1821 to pump water from a well into the canal. Even then the engine was nearly a century old, having been built for a colliery near Nuneaton in 1725. The engine remained in situ at Hawkesbury until 1963; this year was the three hundredth anniversary of Newcomen's

birth, and it was dismantled and re-erected at Dartmouth, Devon, his birthplace, as a memorial. The engine house, which formerly contained both this and a later beam engine, still stands beside the canal, a prominent landmark (grid reference SP 363847).

The most notable canalside landmark of this type, however, is undoubtedly Crofton Pumping Station, close to the Kennet & Avon Canal (grid reference SU 262623). A pump was needed at Crofton to pump water from Wilton Water, a natural reservoir, up into a leat by which it flows to the summit pound of the canal. This commences three quarters of a mile to

the west. Two Boulton & Watt beam engines were installed, one in 1809 and the other in 1812; the 1809 engine was replaced by a new beam engine in 1844. Both these engines were then used regularly until 1952, and the 1812 engine was steamed as late as 1958. Then the chimney of the pumping station was pronounced unsafe and shortened by twenty feet: this reduced the draught so much that the boilers would no longer steam, and electric pumps were installed.

The beam engine installation remained intact, however, and in 1968 was purchased for a nominal sum—about £100—from British Waterways Board by the Crofton Society. This is a branch of the Kennet & Avon Canal Association, the main object of the latter being to reopen the canal throughout for navigation.

Since then both engines have been restored, largely by voluntary labour and with money raised from voluntary sources. They are steamed and operated over weekends at approximately

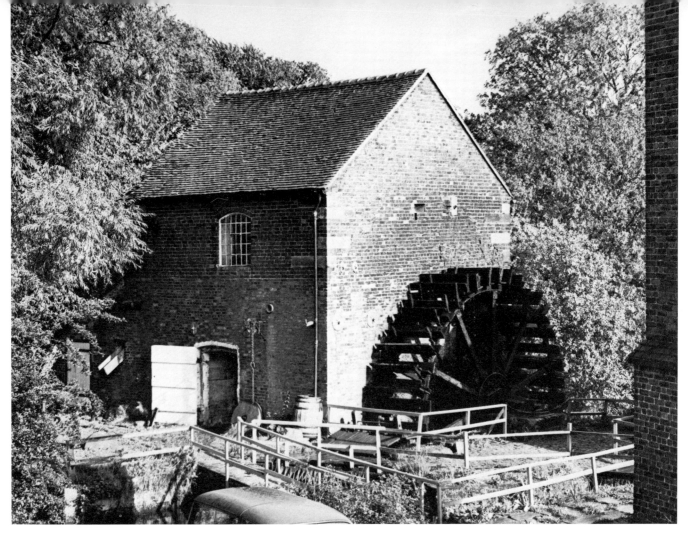

9/16 (top) Cheddleton Flint Mill - the North Mill is seen here from the bank of the Caldon Canal - is, with its rural surroundings, probably typical of the industries served by early canals.

9/17 (above) Claverton Pump House (grid reference ST 792643) contains the Kennet & Avon Canal's unique water-wheel driven pump to supply water to the canal. The lower part of the building to the right contains the wheel, the taller part on the left the pump.

9/18 Knighton milk factory is still a familiar sight beside the main line of the Shropshire Union Canal. It is seen here in 1925: canal boats were used to collect fresh milk from canalside loading points, and for transport of chocolate crumb to Cadbury's Bournville factory, beside the Worcester & Birmingham Canal. The factory now makes Marvel.

monthly intervals throughout the summer, and the public are then admitted on payment of an entrance fee. On other Sundays, when the engines are not in steam, admission is free.

The interior of the pumping station has the usual characteristics of a beam engine house, in which the house itself, divided up into a series of rooms by a central brick wall, forms in effect the frames of the engines. The engines have Watt's parallel link motion and operate at a steam pressure of 20 lb/sq. in.; the cylinder of the 1812 engine has a bore of 42 inches and a stroke of 7 feet 9 inches. The beam of each engine is about 27 feet long. Both operate at 11 strokes a minute and between them can pump 6 million gallons of water in 24 hours. Steam is supplied by Lancashire boilers built by

the GWR at Swindon.

The 1812 engine is, probably, the world's oldest working beam engine.

Following the successful restoration of Crofton, Leawood Pumping Station beside the Cromford Canal has been restored jointly by Derbyshire County Council and the Cromford Canal Society. It contains a single 1849-built beam engine, the purpose of which was to pump water from the River Derwent up to the canal. A horse-drawn trip boat operates between Cromford Wharf, the terminus of the canal, and the pump house. During 1978 this was open to visitors, but not in steam.

Elsewhere reminders of former steam pumping stations, now gone, remain in the names of canal arms built to serve them. Old Engine House Arm, which leads off the Oxford Canal

part way up Napton Locks, is one of them; Engine Branch, which leads off the BCN old main line at the top of Smethwick Locks and crosses the new main line by a handsome aqueduct, is another.

As if Crofton were not enough, the Kennet & Avon has at Claverton, near Bath (grid reference ST 792643) a water-wheel driven pumping station which is unusual, at least so far as canal water supply is concerned. It was designed by John Rennie and started work in 1813; with slight modifications down the years it ran until 1952 when a log jammed in the mechanism and the main drive wheel was stripped of some of its wooden teeth. The pumping station then lay derelict until 1969.

In that year engineering students from Bath University began restoration work. The K & A trust continued it from 1972 onwards, as a voluntary project, and the restored pump house was open to the public in 1978. It operates, like Crofton, over weekends at roughly monthly intervals.

186

The water-wheel is undershot, and in the form of two wheels each 15 feet 6 inches diameter and 11 feet 6 inches wide, on a common shaft. This drives, through gearing, a crankshaft to which are attached two upright 15-feet long connecting rods which in turn drive two overhead rocking beams, of beam engine proportions. The opposite ends of these are connected to the actual pumps. With the water-wheel revolving at 4 rpm, the pumps deliver water at the rate of 77,000 gallons an hour to the canal, which is 53 feet higher.

Steam pumps were occasionally replaced by diesel, as happened at Knowle Locks, where water was pumped back up the flight. As a consequence of the Grand Union widening scheme, a 50 bhp Bolinder two-cylinder semi-diesel engine, second hand from a fishing boat, was installed. Brought back into service in 1974, after fifteen years of disuse, to cope with a water shortage, it awoke the echoes for miles around (literally) and was hastily retired to the Black Country Museum. An electric pump was installed in its place. The Bolinder has recently been moved to Stoke Bruerne.

Canal-served industries

The continuing presence of coal merchants' yards on canal wharves no longer served by boat has already been mentioned, and timber merchants are equally tenacious of their sites. I noticed one recently beside the canal basin at Ripon, disused for navigation since 1892, and they are not rare alongside other canals which remain navigable even though no longer used for trade. The remains of canalside limekilns are also widespread; for example at Dudley, where extensive kilns are being restored as part of the Black Country Museum, at Goytre on the Brecon & Abergavenny and Tiverton on the Grand Western.

On the other hand the canal-served factory, of any antiquity, is becoming rare, though many a bricked-up roving bridge on the BCN shows where once a private arm led off the main canal. Cheddleton Flint Mill (grid reference SJ 972524) beside the Caldon Canal has, I think, the appearance of the sort of industry which canals were built to serve. The mill, or rather mills since there are two of them (one built, probably, by James Brindley) are maintained voluntarily by the Cheddleton Flint Mill Industrial Heritage Trust. Ground flint, which they used to produce, was an essential ingredient of fine china produced in the potteries not far away, so raw materials were delivered to the mill by canal, and its products distributed the same way.

Narrow boat *Vienna*, preserved by the trust, is moored at its wharf to represent the transport used by the mill. The mill itself—a huddle of squat brick buildings, water-wheels slowly turning, set with a few other industries in the midst of the countryside, presents a scene which must have been familiar to early canal boatmen. Today it is open to the public at weekends.

Cheddleton mills were there before the canal; many later industries were sited beside the canal to use it for transport. I have already mentioned in chapter eight how, as late as 1911, Cadbury Bros. Ltd set up a canalside factory at Knighton, beside the Shropshire Union, to collect and condense milk and convert it into chocolate crumb; this was followed by another in 1916, at Frampton, Glos, alongside the Gloucester & Sharpness Canal.

Both of these factories, and the main factory at Bournville beside the Worcester & Birmingham Canal, are still familiar sights to present-day canallers. Their roofed-over quays remain, though it is some years since canal transport was used. Formerly, though, both at Knighton and Frampton, milk in churns was carried by boat from simple collecting stations along the canal to the factories. The main use made of canals, however, was carriage of goods between the out-factories and Bournville, and delivery of fuel.

These two examples of canalside industries, one early, the other late, must stand for the many others which once existed and of which traces, if one seeks diligently, may still be found.

9/19 This mill on the Macclesfield Canal at Marple is perhaps the ultimate in canal-located industry. It still retains an installation for unloading boats, although commercial traffic ceased about thirty years ago.

CHAPTER 10

SOME LOCALITIES WORTH VISITING

Worsley and Barton

Regrettably some planner routed the M62 motorway within a few hundred yards of Worsley village; but the canal here is approaching the entrance to the mine, and is in a shallow cutting below the level of the ground: this and intervening houses and trees render it still a surprisingly quiet mooring, while the motorway interchange does provide an easy means of access by car.

Worsley is one of the few places I have visited which actually appreciates its canal. Pleasant moorings for passing boats are provided beneath the trees at the centre of the village, and a plaque reminds visitors that the transport revolution started here. The Bridgewater Canal, it rightly says, was of 'major importance in the industrialisation of Britain'.

That being so, it is piquant that Worsley, despite being part of the Manchester conurbation, maintains the appearance of a village, or has perhaps reverted to it. It has black-and-white houses, village green and all. The consequences of early industrialisation here lie in the sad grey industrial dereliction which covers so much else of the north-west.

Worsley too is a reminder that, though eighteenth century industrialisation had its appalling side (economics that gave too great a share of the rewards to the owners and too little to the employees, and employment of young children as labourers in coal-mines—there were many at Worsley), yet progress as ever means a step back for every two steps forward. Today's large industries, both nationalised and 'private', are run by faceless bureaucracy: but the Duke of Bridgewater's Worsley estates, and the industries upon them, were run by the Duke, whose residence was Worsley Old Hall. He knew his people, and they

knew him; and by the standards of the time he treated them well. Today the Old Hall is isolated from the rest of the village, beyond the motorway.

Worsley Delph, when I visited it in 1978, appeared as a particularly unlikely industrial site. It can be viewed well from bridges on the A572 that cross over the waterway near the centre of the village (grid reference SD 748005), and is a deepish sandstone quarry cut back into the hillside, surrounded by trees and bushes; the vertical rock faces are half-concealed by ivy tumbling from above, and other foliage takes root where it can. Two very low-roofed tunnels, looking like the drainage tunnels as which they were largely built, emerge from the far rock face at its base, and an isthmus between them projects into the canal basin which otherwise occupies the quarry floor. It is not easy to appreciate that this bosky grotto was the entrance to mines which extended for four miles to the north and were once served by a total of forty six miles of underground canal on three levels, with an inclined plane between two of them. Here was a flourishing industry for nearly ninety years: then, after a decline, the underground canals were last used for transport in 1887. But brownish-yellow water still pours from them to feed the main canal, and the remains of sluices, which once adjusted currents to suit movement of boats inside the mine, are still to be seen outside the tunnel mouths. Most interesting of all, at the time of my visit, was a real starvationer boat, moored to one of the rock faces and bringing the whole scene into focus. In fact, it was a survivor of those used, since mining ceased, for tunnel maintenance.

The canal at Worsley takes on the layout of a capital T, the line from Manchester and the South forming the upright, the approach to the mines the right-hand arm, and the commencement of the line to Leigh and the North the left-hand. This, the Leigh branch

of the Bridgewater Canal, was opened throughout about 1799, but its commencement at Worsley is much older, for it originated in 1759 as part of the line heading for Hollins Ferry. This was never completed but the section that was built was used for local transport before being, in part, incorporated into the Leigh branch.

Between the two branches, at the head of the T, stands the fine old half-timbered building called the Packet House, where tickets for the packet boats were sold. Beside it is the wide flight of stone steps which passengers used to descend to join the boats for Manchester. And a hundred yards or so in that direction, on the east bank of the canal, stands a little ornamented boat house. This once housed a state barge built by the Earl of Ellesmere, the Duke's successor, for the occasion of Queen Victoria's visit to Worsley in 1851. She came by train as far as Patricroft and from there, to Worsley, along the canal.

From Worsley it is two and a half miles down the canal to Barton. When the Manchester Ship Canal was built it occupied, at that point, the course of the Mersey & Irwell Navigation. This meant that Brindley's aqueduct had to be demolished, and replaced by a structure able to let large ships pass along the ship canal. The solution adopted by Leader Williams, the ship canal company's engineer, was as adventurous as Brindley's: an aqueduct with a span which swings open, like a swing bridge, to allow ships to pass. Stop gates seal off the Bridgewater Canal each end of the aqueduct and both ends of the span, enabling it to be swung full of water, with consequent saving of both water and time.

The swing aqueduct was built alongside and to the east of the original aqueduct, and the Bridgewater Canal diverted across it. Illustration number 10/2 shows the swing aqueduct under construction, with arches of the original aqueduct beyond. Building the

swing aqueduct before the old one was demolished must have kept interruption of Bridgewater Canal traffic to a minimum, but it does suggest ample confidence in the design of the new aqueduct, despite its originality, for it would have been impossible to run trials, swinging it open and shut, until after the old aqueduct had been demolished, that is, to say until the trough of the new one had been in use for some time.

To say that the old aqueduct was demolished is not wholly accurate. Certainly the three spans which crossed the Irwell, which would in today's terms have been called the aqueduct, were demolished. The original aqueduct, however, is often said to have been 200 yards long, and since the existing aqueduct has a span of 235 feet and the three cross-river arches of the original were shorter still, this suggests that contemporaries, in speaking of Brindley's aqueduct, referred not only

to the section which actually crossed the river, but also to another arch across Barton Lane to the north, and a stone-built embankment which linked the two. Their total length would approximate to 200 yards. Early engravings, too, often show the whole of this structure.

When the canal was diverted, a new aqueduct was built across Barton Lane, and a new embankment to link it to the swing aqueduct. So the whole of Brindley's structure went out of use. But the central masonry embankment section of it still stands where it always did and, viewed from Redclyffe Road which runs parallel, looks like a rampart. Furthermore, although the original arch over Barton Lane was taken down, one of its faces was preserved and re-erected, turned through ninety degrees, beside the road on the site of its original northern abutment. There it remains, a memorial to England's first canal aqueduct.

10/1 (above) The entrances to the underground canals at Worsley mine, with starvationers in the foreground. The scene is very similar today.

10/2 (overleaf) Barton Swing Aqueduct under construction with, beyond it, Brindley's original aqueduct still in place.

10/3 (above) Barton Swing Aqueduct: it is unlikely
that it was ever regular practice to swing the
aqueduct with a boat in the trough, let alone a horse
on the towpath – probably this early scene was
trumped-up for the benefit of the photographer.

10/4 (opposite) The Horseshoe Falls in the Dee at
Llantysilio are in fact a weir to hold up water for
supplying the Shropshire Union Canal. The intake is
out of the picture to the left.

The swing aqueduct is a worthy successor. With its grey painted girders it swings silently, swiftly for what it is, by hydraulic power, and does not leak much. The weight of the moving structure is 1,450 tons, which includes 800 tons of water. While it moves, a central hydraulic press thrusts upward to relieve its roller bearings of about 900 tons of the total weight. The aqueduct is mounted on an island in the ship canal which it shares with a road swing bridge and the control tower for both of them. When the swing span is closed across the ship canal, the joints between its trough and the Bridgewater Canal are kept watertight by rubber-faced wedges operated by hydraulic rams. All in all it is a grand piece of Victorian engineering still functioning well in its intended manner.

Llangollen, Pontcysyllte and Chirk

A pile of stop planks and a warning notice in the name of the Shropshire Union Railway & Canal Co. seem out of place among the sylvan surroundings of the Horseshoe Falls, in the Dee Valley at Llantysilio. But both are concerned with the water intake of the Ellesmere Canal, and the falls themselves (grid reference SJ 197433) are artificial, a weir built to maintain the water level of the river, for canal feed.

As the valley deepens and the river descends, the canal maintains its level and by Llangollen wharf (grid reference SJ 215423) is already on a shelf on the hillside high above the town, so that a sign below indicating boat trips points incongruously up a steep hill. The horse-drawn passenger trip boats here have operated since 1884 and remain the best way to see the canal above Llangollen, for it is exceptionally narrow and the last winding hole is a few hundred yards beyond the wharf. The horses themselves are stabled in the rear of the wharf building: a unique example of canal stables in full and continuing use.

The wharf building itself has become an excellent canal exhibition centre: as visitors move through the exhibition, the story of canals is progressively revealed by means of slide displays, sound effects, photographs, models, artefacts and striking trompe-

10/5 (above) Canalside cast iron notices often recall names of long-gone companies. These two examples are at Llangollen Wharf; the SUR & C Co was merged with the London & North Western Railway in 1922.

10/6 (right) Pontcysyllte Aqueduct: the iron trough and ribs in close-up, looking north. Bases of temporary supports used during 1975 repairs can be seen in the foreground.

l'oeil murals that at one point create the illusion that one is standing on a canalside wharf busy with trade.

The canal continues on its hillside shelf, very narrow as befits a canal built principally as a feeder, and with a noticeable current, for it is now used for general water supply as well as canal feed. At grid reference SJ 241424 an otherwise ordinary humped bridge has a small additional arch: through this there once passed a tramroad connecting with the canal. At Trevor the canal debouches under a bridge into what was intended as the Ellesmere Canal's main line—to the left, however, it extends for only a few hundred yards to former wharves, now

a hire base. Beyond this point the canal was never completed. Immediately to the right is Pontcysyllte Aqueduct (grid reference SJ 271420), the highest and largest canal aqueduct in the British Isles and the greatest monument of the canal era. Its history and construction are described in chapter seven.

The aqueduct is so large that it is best viewed, initially, from a distance. People coming down the canal from Llangollen will already have had a fine view of it from the canal a quarter of a mile back; it can also be seen well, from the canal or the A5 road, at grid reference SJ 276412. From these viewpoints the full extent of its 1,007 feet

length can be appreciated, and so can the massiveness of the southern approach embankment—at the time of its construction the largest canal earthwork then built.

To see the full height of the aqueduct (127 feet) there is a good view from the road bridge over the Dee at grid reference SJ 268421. From here the highest spans can be seen carrying the canal across the river, but trees conceal the full length of the structure. To cross the aqueduct by boat is a remarkable experience, for though there is a towpath and railings along the east side of the channel, on the opposite side is only the slender edge of the iron trough, projecting a foot or so above

water level. So looking over the side of the boat, one looks directly down to the valley floor 100 or more feet below: the nearest approach to ballooning that it is possible to achieve by boat.

At Froncysyllte, at the south end of the southern approach embankment, there is a fine example of a typical Ellesmere Canal wooden drawbridge (grid reference SJ 272413).

A childhood experience which impressed itself markedly on my mind was that of eating lunch in the dining car of an express train passing through the Severn Tunnel. It was an experience of which I was immediately reminded recently when, after making an early start during a cruise down the Ellesmere Canal, I found myself relieved from the tiller and able to go down into the cabin for breakfast while we passed through the tunnel at Chirk. There was the same contrast between the security and domesticity of a meal in a brightly lit saloon and the darkness beyond the windows suggestive alternately of cosiness and peril. Only the speed was different.

Chirk was one of the first tunnels of any great length to be built with a towpath. It is in fact 459 yards long and the width of the channel is insufficient for boats to pass one another. At its southern end the canal emerges into a wide pool (grid reference SJ 286375) where boats may moor or pass, and then immediately narrows once more as it goes onto Chirk Aqueduct.

Chirk Aqueduct is 600 feet long, 70 feet high at the highest, and has ten arches. Its story, and a description of its curious part-masonry, part-iron construction, are given in chapter seven. Chirk Aqueduct is overshadowed figuratively by Pontcysyllte to the north and literally by a later railway viaduct alongside. But although Pontcysyllte is the more spectacular of the two aqueducts, its spans though extensive are black and sombre: Chirk, built of lighter stone, and with the tunnel mouth contiguous, seems to me the more dramatic.

Stoke Bruerne and Blisworth Hill

10/7 (below) Stoke Bruerne in the 1950s. The empty warehouse building at the far end of the row of houses on the right was later to become the Waterways Museum. A pair of coal-laden ex-G.U.C.C. Co. narrow boats is descending the lock, which dates from the late 1830s when locks were duplicated to relieve congestion. Remains of the original lock are on the left.

10/8 (opposite) Stoke Bruerne, 1978. Pleasure craft have replaced trading boats, and the Waterways Museum and its visitors have arrived.

The canal and its surroundings in the vicinity of Stoke Bruerne and Blisworth are of great interest, for not only does the canal village of Stoke Bruerne contain the national Waterways Museum, but there is also Blisworth Tunnel a little to the north of it. With a length of 3,075 yards this is the longest canal tunnel now in full use as such (if one discounts Dudley Tunnel as comprising two tunnels with a short open-air section between them and navigability restricted by limitations on headroom and power).

Stoke Bruerne was important to canal people long before the museum was established. It is the only village through which the Grand Junction, later Grand Union, Canal passes for several miles. It lies at the head of a flight of locks, and many of its buildings line both sides of the canal in the short space between the top lock and the approach cutting to the tunnel. It was therefore a convenient and popular place to moor and obtain supplies. Hollingshead passed through 'Stoke Brewin' in 1858 during his voyage from London to Birmingham by fly boat described in *On The Canal*. He took a jaundiced view—he was unable to obtain meat at the butcher or food in the canal-side tavern—but then he and his companions were tired and hungry. A. P. Herbert described Stoke Bruerne in the late 1920s in happier terms in *The Water Gipsies*; and though the story is fiction, the description of Stoke Bruerne has the ring of fact. He called it a very pretty little place, compact and sleepy and altogether English; the Boat Inn stood on one side of the lock and the chandler's and grocer's on the other, and many horse boats tied up at Stoke Bruerne

for the night to go through the tunnel with the first tug in the morning.

The Boat Inn is built of local stone, like many old buildings in this village, and thatched. The building pre-dates the canal, and the inn is still busy, having outlived the commercial traffic on which it flourished for so long. It stands on the west side of the canal, beside the top locks of the flight. The Stoke Bruerne flight was duplicated in 1835, although use of paired locks only lasted a few years, after which one of each pair went out of use. At the top lock it is the newer lock of the pair which is now in use, on the eastern side of the canal. The chamber of the original lock, dating from *c.* 1805, has become an open-air section of the Waterways Museum. In it has been installed a barge-weighing machine which originated on the Glamorganshire Canal at Cardiff. This was used to measure the tonnage carried on a boat, and so to enable the toll to be calculated: a loaded boat, of which the

weight empty was already known, was floated into the cradle of the machine and water was drained away, leaving it resting on a platform comparable to that of a weighbridge.

Preserved in situ on the weighing machine is the horse-drawn narrow boat *Northwich*. She was built in 1898 for Fellows Morton & Clayton Ltd and has been restored to her original colours. As well as the usual cabin at the stern she has a small fore-cabin in the bows, a feature of boats inhabited by large families. By 1949, when the FMC fleet was acquired by the British Transport Commission, *Northwich* was employed in FMC's North Western Division, as motor-boat butty on the Shropshire Union Canal, where she carried refuse.

Installed at the south end of the lock chamber, which was narrowed to fit them, are a pair of narrow-lock gates made, unusually, of cast iron. These originated on the Montgomeryshire Canal at Welshpool.

Twin humped bridges survive at the tails of these locks, though the original one now spans dry land only. Both bridges are quite sharply skewed, and combine both brick and stone construction: arches are built of brick, the rest of the bridges of stone, except that the new bridge has brick parapets. Both bridges are now whitewashed.

Along the east bank of the canal above the lock two rows of cottages of successively increasing height culminate in the fine three-storey stone building which houses the Waterways Museum. It was once a grain mill which owed its location to availability of canal transport, and was powered by a beam engine housed in a building immediately adjoining, on the site now occupied by the Waterways Museum Shop. This is of recent construction, but the roof-line of the earlier building, which was taller, can still be distinguished on the northern wall of the museum building proper. Old photographs of Stoke Bruerne show that a

tall chimney to the engine house was once a distinctive feature of the village. In recent years milling had ceased and the main building was used as a grain warehouse only.

By the early 1960s it was disused. The newly-formed British Waterways Board restored it and opened in it the Waterways Museum in 1963. It would have been difficult to find a more appropriate place: a canalside building which owes its existence to the canal, in a location where visitors can see not only the museum exhibits, but also, within a few hundred yards, boats passing through locks and entering and leaving a long tunnel. Furthermore it is easy of access, whether by canal or road, for the nearest M1 intersection (number 15) is but four miles away.

The private collection of lock-keeper and former Number One Jack James formed, on loan, the nucleus of the initial exhibits; the museum's collection has since expanded, acquiring amongst other material the canal exhibits from the former British Transport Museum at Clapham, London, and it is now the definitive collection of British canal artefacts.

The most striking and in many ways the most interesting of these is the narrow boat cabin. It was never easy for strangers to see inside the cabin of a family boat, for to peer in was as offensive as to peer through anyone else's front door. Here, however, one can peer to one's heart's content. The structure is an accurate replica of a narrow boat cabin and stern end, and was built for the museum in 1962–3 by the traditional boat yard at Braunston, formerly Nurser's, but by then owned and operated by Blue Line Cruisers Ltd. It bears the name *Sunny Valley*, originally adopted for a boat appearing in the 1940s film *Painted Boats* and familiar in the early 1960s from the jacket illustration of L. T. C. Rolt's *The Inland Waterways of England*. The material is mostly hardboard, but measurements and shape are absolutely exact — the replica was built by the yard's foreman boatbuilder — and the paintwork and furnishings are real. The interior of the cabin was arranged by Jack James in the traditional manner, and the whole is 'decorated and furnished as it would have been by the proudest boatman of the old school', as the museum guide book expresses it. The yellow-brown grained paintwork

of the interior is largely hidden behind a profusion of crocheted lacework, 'lace' plates and gleaming brasses, an oil lamp on a bracket and a large black stove, while exposed panels display fine and colourful rose-and-castle paintings.

That so many things should be crowded together in this way is I suppose inevitable when so small a cabin provided living space for a family, but nevertheless the style and manner in which it was done always puts me in mind of the way in which our Victorian ancestors cluttered up their living rooms with objects in confusing quantity, and I suggest that the narrow boat interior indicates a survival on the water of commonplace Victorian customs long after they had been superseded elsewhere. The same seems true of other features of canal life, such as boat people's costumes, of which there is a good display in the museum. Old boatwomen in the 1930s were still wearing bonnets of a type fashionable in the 1890s.

Near the boat cabin stands a full-size replica horse, wearing genuine boat-horse harness, with multi-coloured bobbins on the traces to prevent their chafing his sides, and a crocheted cap to keep flies away from his ears. The harness came in part from the BCN and in part from the Leeds & Liverpool. Nearby too can be seen the horse's rival: a 15 bhp single-cylinder Bolinder semi-diesel engine. This example was made about 1928. The built-in blowlamp used to pre-heat the hot bulb can be seen, and so can the arrangement of exposed rods, levers and tappets provided so that the boat could be put astern by reversing the engine's direction of rotation.

There are numerous other exhibits. They include an excellent collection of early engravings and old photographs of canals, pictures of one sort or another of many of their principal engineering works past and present, and a large-scale model of the Anderton Lift. There are collections of painted canalware and canalside notices. There are curiosities like part of an enormous semi-circular brush which, boat-mounted, was used to remove soot deposited on the interior of Blisworth Tunnel by steam tugs and narrow boats, and sad relics such as the name board from the last canal packet boat *Duchess-Countess* which is accompanied by photographs of and

plans made from her shortly before she was broken up about 1956.

The high ground called Blisworth Hill, a little over a mile to the north of Stoke Bruerne, forms the watershed between the valleys of Great Ouse and Nene and provided the builders of the Grand Junction Canal with their most formidable obstacle. The first attempt to bore a tunnel through it during the years 1793 to 1796 ended in failure when excessive water flooded the workings. By 1796 the northern section of the canal was open from Braunston as far as Blisworth, and the company built a toll-road onwards over the hill. In 1800, when the southern section of the canal had reached the point that was to become the foot of Stoke Bruerne locks, the road was superseded by a horse tram-road built to connect the two ends of the canal. Meanwhile the canal company had considered crossing the hill permanently by flights of locks, but eventually decided to bore another tunnel on a new alignment. First, however, a heading or smaller tunnel was made beneath the line of the main tunnel to drain away excess water. The tunnel (wide, but without a towpath) was eventually opened in 1805 and with it the canal's main line was complete. The nature of the ground through which the tunnel was bored — wet, and liable to move and distort the tunnel lining — has subsequently caused problems. During the latest instance, when the tunnel had to be closed for repairs from May until October 1977, it was found that part of the invert had been thrust upwards by as much as one metre.

Visiting the area in 1978 I was able to find traces of all three earlier attempts to cross the hill, of construction of the present tunnel, and of past methods of working traffic through it. The horse tramroad must have run close to, if not actually across, the site of the museum; its course cannot be traced there but could be clearly seen (by plunging through the undergrowth) a little to the north, where it appears as a ledge high up on the east side of the tunnel approach cutting. Here the tramroad, climbing steeply, must have passed through a shallow cutting of which the west side was removed when the deeper canal cutting was excavated.

At canal level, shortly before the tunnel mouth, the horse path diverges

10/9 The north approach cutting to Blisworth Tunnel, Grand Union Canal, at grid reference SP 726531. The ledge in the foreground marks the course of the temporary tramroad which preceded the tunnel, and the iron boundary post marked GJC indicates the limit of the Grand Junction Canal Company's property.

from the towpath to ascend the hill. Where it does so there stands a brick-built shed that formerly provided temporary stabling for boat horses: having walked over the hill, they sometimes arrived long before their boats if the latter had had to wait for some time for the tunnel tug's next run. Another building close to the tunnel mouth was once a base for the tugs and used to contain coal store and work-shop.

Much of the minor road which leads from Stoke Bruerne to Blisworth clos-ely follows the course of the tunnel below, and this section was originally the canal company's toll road. The horse path from the southern tunnel mouth joins it, and from it many air shafts can be seen. For safety, and to increase the draught which extracts smoke and fumes, their mouths are enclosed in cylindrical brick structures which resemble squat, wide chimneys; around them are mounds of spoil excavated from the tunnel. Where the

10/10 This tramroad wagon of the Stratford & Moreton Railway is preserved on the course of its original line close to Stratford canal basin.

road runs close to the northern en-trance to the tunnel, an inclined horse path leads down again from road to towing path; and near the tunnel portal itself stands another steam-tug coal store. Back on the road, and close to the tunnel's northern entrance, a bridle path leaves the road and from it at grid reference SP 729527 a curving embankment can be seen. This carried the horse tramroad across extensive depressions in the ground which are, I believe, traces of the first attempt to excavate a tunnel. The tramroad is said to have been double track, but the narrowness in places of its surviving formation leads me to believe that parts of it at least were single.

As at the south end of the tunnel, so at the north the canal's S-curved approach cutting closely parallels the course of the tramroad, which here appears as a ledge a little below ground level on the western side of the cutting. It seems to have become a public footpath: I followed it while the cut-ting grew shallower and a series of cast iron boundary markers, lettered *GJCCo* and set well back from the edge of the cutting, marked the course of the tramroad, and presumably even today the boundary of **BWB** property. Blis-worth wharf, where the tramroad joined the canal, could be seen in the distance.

Stratford-upon-Avon and its canal

People visit Stratford in enormous numbers, of course, and a great many of them visit the Shakespeare Mem-orial Theatre and the riverside gardens that lie between it and the main-road bridge across the Avon. It is another question whether many of them stop to ponder why, in the midst of the munici-pal lawns and flower beds, there is a large sheet of water with long narrow boats with cabins moored upon it, and with a lock up from the river at one end and an insignificant-looking bridge beneath the road at the other; why, in an area obviously given over to recreation and amenity, a prime triangular chunk of land between gardens, road and river should be occupied by the pre-mises of a timber merchant which have all the appearances of being old-established; or why, should they wish to cross the river to visit the park beyond, the pathway by which they do so is carried by an attractive many-arched brick bridge, substantial yet narrow, which strides across the river straight and level and purposefully . . . to no obvious purpose.

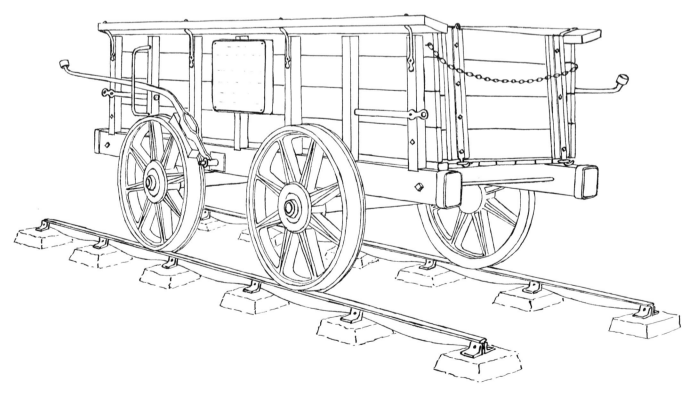

10/11 *Stratford Canal: Wootton Wawen Aqueduct carries the canal over the main Stratford-Birmingham road.*

If however they find a curious and primitive railway wagon which stands on a short length of track beside the pathway leading to the bridge, and next to the timber merchant, and if they read the plaque attached to it, they learn some, but not all, of the story. The wagon is an original relic of the Stratford & Moreton Railway, mentioned in chapter five. It is typical of the type of vehicle used on tramroads connected with canals: narrow wooden body set between the wheels, no springs, wooden main frames extended as unsprung buffers, its brake a wooden block actuated by an iron lever. Very few such vehicles still exist: this one not only survives but is preserved in situ, for it stands either on or very close to the original course of the line, and on a section of original track: iron rails and iron chairs mounted on stone sleeper blocks. Much of this, and a potted history of the tramroad, the visitor can learn from the plaque or by observation. Of the wider implications of the presence of

the wagon on this particular spot, beside river and canal basin, the plaque is silent.

For of course the sheet of water a few yards away with the boats on it is the basin of the Stratford Canal and the whole area now given over to Shakespeare worship and tourism was at one time the transport centre of Stratford, where the canal, the Avon navigation and the Stratford & Moreton Railway all met. It was an area of commerce, of wharves and basins, tramroad sidings and turntables, bridges and warehouses. The Shakespeare Memorial Theatre itself occupies the site of a second canal basin: part of the arm to it is still extant, leading out of the westernmost corner of the basin which survives. The timber merchant is the sole remaining example of many such premises in the area served by water and tramroad, and that hand-

some bridge which now carries the footpath was built to take the tramroad across the river on the first stage of its route to Moreton.

Outside Stratford the course of this line is still easy to find, although it was last used in the early years of this century and the track was taken up during the First World War. It leaves Stratford on a substantial embankment, which now carries a footpath, and includes a large underbridge which, with massive abutments and red brick parapets, has a distinct canal-era look about it. This is located where the course of the tramroad comes alongside the A34 main road. It follows this road for several miles, showing up as a widening of the grass verge which becomes a shallow cutting or a low embankment wherever the road climbs or falls too steeply.

On the Stratford Canal itself, the

southern section from Lapworth to Stratford is a pleasant example of a canal built during the middle years of canal construction, which also has several distinctive features. Although authorised in 1793, it was not built until the period 1813–16. The story of its later decline under railway ownership, to the point at which it was all but derelict, and subsequent recovery, largely by volunteer labour under the ownership of the National Trust, is an epic one which I have mentioned briefly in chapter six and told fully in *Waterways Restored*.

The canal has the typical ingredients of an English rural canal. It is narrow (only the lock from the river up into Stratford basin is wide, to admit river barges) and it winds its way, without being excessively circuitous, through a countryside which is extremely pleasant without being outstanding. Most of its locks, of which there are rather a lot, are grouped, not too closely, in flights at each end.

There are three distinctive features. The first of these is the three iron trough aqueducts. A small one crosses a stream at Yarningale and dates from 1834, as it replaced an earlier one washed away by a flood caused by a breach in the Warwick & Birmingham Canal which crosses the same stream further up. Two larger ones date from construction of the canal, one across the main road at Wootton Wawen and the other, the largest, at Edstone which not only crosses over a stream and a road but has had a railway routed beneath it also. There is more about these in chapter seven.

The second distinctive feature is the 'split' bridges—that is to say, bridges of which the arch appears to be split, so that tow-ropes could pass through without being detached from horse or boat (there is no towpath beneath these bridges). In reality they comprise two substantial iron brackets supported from the abutments, with a gap of an inch or so between them. Such bridges were used on other canals built before the southern Stratford; the Trent & Mersey and the Stourbridge are examples, which used them in the vicinity of locks for access; but only on

the southern Stratford were they adopted throughout the line with such enthusiasm.

The other feature of the southern Stratford, one which is unique, is the several surviving lock cottages with barrel-vault roofs of semi-cylindrical brickwork mentioned in chapter nine. They were built, it appears, by men who knew how to make bridges but were unfamiliar with roofs: so for a cottage they built a longish bridge arch, walled the ends and inserted windows.

Birmingham, new lines and old

The Birmingham new main line comes as a revelation. It is approached, from the east, either by the level and surprisingly leafy Worcester & Birmingham Canal to Gas Street basin; or up the successive and seemingly interminable flights of narrow locks, hemmed in and hidden behind factories and beneath railway bridges, of the Birmingham & Fazeley. Either way there is nothing to prepare one for the new main line: a broad watery highway which leads, by graceful curve and lengthy straight, along deep cuttings and beneath high bridges which represent the canal builder's art at its zenith, through one of England's greatest and busiest conurbations; and yet it is now almost deserted.

Its story is this. The original Birmingham Canal, a contour canal, was authorised in 1768 and its first section, from Birmingham to Wednesbury was opened in 1769. The attraction of Wednesbury was to bring coal thence to Birmingham. Then, diverging from the first section at Spon Lane, Smethwick, the canal was extended to Wolverhampton and Aldersley, its junction with the Staffordshire & Worcestershire Canal, which was reached in 1772.

To describe the canal simply as a

10/12 The deck of a 'split' bridge, typical of the Stratford Canal, was built in two sections, bracketed from the abutments, with a space between for tow-ropes.

contour canal is an understatement: like the Oxford Canal, it was exceptionally circuitous—Simcock, one of Brindley's assistants, had much to do with both. As the area through which it passed grew in prosperity, so the traffic of the canal grew likewise. Other canals were built to connect with it, and it became increasingly congested with boats. The first attempt to improve it was made at Smethwick. John Smeaton was the engineer. The original canal passed over its summit here by a short pound approached by six locks on either side: the line to Aldersley diverged from the Wednesbury line between the third and fourth locks on the Wednesbury side. So a new length of canal was excavated, in a cutting, to replace the original summit pound and eliminate the top three of the locks at each end of it. This meant that the canal became level from Smethwick to Wolverhampton; the three locks remaining on the Birmingham side of the summit were duplicated.

Little else was done until 1824. Then the Birmingham Canal Navigations company, which owned the original canal and others connected with it, commenced a series of improvements. It was prompted to do so partly by proposals for the Birmingham & Liverpool Junction Canal which was likely to bring still more traffic, and partly by proposals for construction of railways which, in the absence of improvements, were likely to reduce it. At any rate the company began its improvements, with Telford as engineer in the early stages, and these—diversions, straightenings, shortenings, duplicate and new lines—were continued at intervals into the 1860s, well into the railway age. By then the company had come, rather loosely, under railway control. The reason why it continued to expand arose from the existence of a great many canalside factories and works which, lacking rail connections, continued to use boats for delivery and collection of goods to and from canal-railway interchange basins. In 1905 there were twenty six such basins in use, and the BCN continued to be busy with short haul traffic for many years.

The Birmingham canals can best be seen by boat, if only because so many of them are still walled in, with few points of access from streets, although the main railway between Birmingham

New Street and Wolverhampton runs close to the canal for much of the distance, and a journey from one station to the other does enable one to see a lot in a little time. However the best way is by boat, starting, as I did recently, from Farmer's Bridge, Birmingham, where the new main line commences.

Telford recorded that on inspecting the original canal he found it 'little better than a crooked ditch with scarcely the appearance of a haling path, the horses frequently sliding and staggering in the water . . . and the entanglement (of lines) at the meeting of boats incessant . . .'. So he laid out the new canal with dual towpaths, one on either side. It leaves Farmer's Bridge by a gentle, flowing right-hand curve, and the first bridge is a foretaste of things to come. Like earlier canal bridges, it is built of red brick, and humped—though the curve of the parapet is so slight as to be scarcely noticeable. But this bridge has twin arches, the pier between them rising from the centre line of the canal; and each arch is wide enough to span not only ample width of water for passage of a boat, but also the adjoining towpath, without interrupting the alignment of the canal bank. No need here for a horse to swerve under a narrow bridge hole or duck to avoid a low arch.

The canal leads on, and soon becomes straight as far as the eye can see. At intervals there are junctions and at least one 'crossroads' where the new line intercepted the old, the loops of which remained in existence for local traffic. At each junction, the canals are spanned by cast iron roving bridges. At intervals, too, come the islands which once carried octagonal toll houses (all now demolished) with gauging 'stops' on either side.

There are two magnificent bridges, Lee Bridge and Winson Green Bridge (grid reference SP 043879). They represent the canal bridge grown up. Each is a brick arch, but high, wide, spanning the full width of canal and towpaths, on a skew so that the courses of streets above are not affected, dwarfing the railway bridges built later alongside, and constructed, according to plates they bear, in MDCCCXXVI.

As built, these bridges had wrought iron open parapets: they were two of the few places where Birmingham canals were open to view rather than

hidden. Later, probably about the time of the First World War, the parapets were raised, filled in and used as advertisement hoardings. More recently, the condition of the ironwork gave cause for concern: so the parapets were taken down and replaced by open handrail. Passers-by once again enjoy the view of the canal.

At Smethwick the old main line diverges to the right—wherever the new line parted company with the old, the junction was laid out to give the new line the straight run. The old line ascends three locks, those built by Smeaton: traces of the original locks remain beside them. The new line enters the deep and curving Galton Cutting, but the two lines run parallel to one another for a mile or so. Telford

Aqueduct—sombre, black, cast iron and handsome—spans the new main line and carries a short branch of the old which once led to a beam engine used to pump water back up the locks to the summit.

The next bridge (grid reference SP 019888) carries Brasshouse Lane (name redolent of canal-age Birmingham!) and from a viewpoint on it, or better still from the railway footbridge adjacent, the story of these canals is revealed. The new main line, straight and wide, runs deep in the cutting far below. To the right, the narrower old main line is seen at a higher level. Farther to the right again, a shelf higher still on the cutting side marks the course of the original summit pound. From here, too, one used to get

a vista down the new main line to Galton Bridge, the greatest bridge of all, a graceful cast iron bridge which spans the cutting at its deepest from brim to brim (grid reference SP 015894). This view is now lost, for to construct a new road part of the cutting was filled in and a new canal tunnel made beneath it. As my boat emerged from this I was wondering what had become of Galton Bridge only, on looking up, to find it, suddenly and unexpectedly, high overhead.

Three quarters of a mile further on, Stewart Aqueduct carries the old main line over the new. It lies in the shadow of the overhead M5 motorway, which here straddles the old line. Beyond, on the right, the three Spon Lane locks

10/13 The three levels of the Birmingham Canal at Galton Cutting, grid reference SP 018889. The position of Brindley's original summit pound is indicated by the ledge high on the cutting side to the right; Smeaton's improved line is below it, and Telford's new main line, wide and straight, below that.

descend from the old line to the new. They are the only working survivors of the twelve locks on the original Birmingham Canal brought into use in 1769: the oldest narrow locks. Some of the locks on the Staffordshire & Worcestershire canal were built earlier, but did not come into use until the following year. At Bromford Junction (grid reference SO 998898) at the tail of Spon Lane locks (more fine cast iron roving bridges) the new main line takes up the course of the original Wednesbury line, though not, I suspect, its alignment, much of which has been straightened. Then, at Pudding Green Junction, it strikes off to the North West, straight as a die, towards Tipton and Dudley. At Tipton the new line joins the old, but beyond again another new section of about one and a half miles, including Coseley Tunnel with dual towpaths, enables boats to avoid more meanderings of the original route, and the rest of the line to

Wolverhampton was improved. These works were completed in the late 1830s.

Little seems to have been done to the flight of twenty one locks leading down from Wolverhampton to Aldersley, and the lowest locks in the flight have now a bucolic appearance. The entrance lock from the Staffs & Worcs is particularly unimposing (though none the less pleasant). Tucked away in a small wood behind a low brick roving bridge, it gives approaching boaters as little hint of what lies above and beyond as do the eastern approaches to Birmingham. But close inspection shows that innumerable feet have worn away the stone treads of steps at the tail of the lock, and countless towropes have deeply abraded a corner of the bridge's brick arch.

It is of equal or greater interest, I have found, to diverge at Pudding Green and head for the Tame Valley Canal. For the first three quarters of a

mile the route is the original Wednesbury line; then one branches off to the left down Ryder's Green locks, along one of the first extensions to the original canal. At the foot of these locks, at Great Bridge, (grid reference SO 978927) is a good example of a railway interchange basin, or what is left of it. No longer in use, it is still in water: some waterside railway sidings remain, but most of the fingers of land jutting out into the basin, which once carried the railway tracks, are now overgrown. Between them, a few old boats disintegrate among the reeds, and the roof which once covered part of the basin and some of the sidings has disappeared.

Not much farther on, a sharp turn to the right leads on to the Tame Valley Canal. This heads back to the east, and joins the Birmingham & Fazeley Canal at Salford below the two flights of locks by which that canal ascends to Farmer's Bridge; and it was built in

part as a diversionary route, for those locks, though extremely busy, were too much built-up on either side to be duplicated. It was opened in 1844.

The Tame Valley Canal makes an interesting contrast with those parts of the new main line which were built fifteen years or so earlier. It does indeed have high embankments and deep cuttings, dual towpaths and also dual stop gates, with a small island between them, at the ends of embankments as a precaution against breaches. It has a few cast-iron roving bridges, plainer than the earlier ones, and bridges of brick with iron beams. It has long straight stretches and occasional curves. But although the civil engineering works are on as great if not greater scale than those of the Birmingham new main line or the Birmingham & Liverpool Junction, the Tame Valley Canal is not laid out with the same flair: it is derivative work, not original. The canals of the 1820s and 1830s represent the peak of the canal engineer's art, and their structures and layout an expression of triumph. By the 1840s, all is wholly utilitarian and, I suspect, the best of engineers had deserted canal work for railways.

The old and new lines of the Oxford canal

One does not, often, chance upon a canal at the precise point at which, though in water, it becomes un-navigable. Standing on the bridge at Stretton (grid reference SP 441809) on a hot day in May 1978, however, I found I had done just that. Looking down the canal, the waterweed grew

10/14 (below) The new main line, right, of the Birmingham Canal Navigations, diverges from the original Wednesbury line at grid reference SO 998898. The tail of the lowest of the three Spon Lane locks can be seen on the left: they were the first narrow locks to be brought into use, in 1769. The cast-iron roving bridges are typical of the BCN in the early 19th century.

10/15 (overleaf) Great Bridge railway interchange basin, seen from the bottom lock at Ryder's Green (grid reference SO 978927). Railway sidings formerly extended along the overgrown promontories so that rail wagons and boats could exchange loads.

less, the overgrowing shrubbery gradually parted, and in the distance narrow boat style cruisers were moored to the bank. The bridge upon which I stood was a typical canal humped bridge, but alongside it a newer, low level bridge had been built to carry road traffic. Beyond it, up the disused canal, lay a typical small country canal wharf. This was Stretton Wharf, half a mile from Stretton under Fosse which village it served, beside a very much older transport route, the Fosse Way, its small warehouse still standing, its canal devoid of boats and inaccessible to them, and its open spaces crowded with parked lorries.

The canal was part of the original line of the northern part of the Oxford Canal. When the improved and shortened line was opened in 1834 it bypassed Stretton Wharf: a quarter-mile of the old line was left in water as a short branch to serve it. Several other village wharves, such as Brinklow, were treated similarly, and these and many other traces of the original line are still to be found. A car is the best means of transport when looking for them, unless, while boating, one is prepared to moor up for longish periods of exploration on foot, and the Ordnance Survey map suggests a long time could be devoted to it whatever the means of conveyance. My own time being short, I was able only to visit a selection of sites.

The state of the old line varies and is even now changing. Beyond Stretton Wharf, it can be seen as an embankment, quite high by contour canal standards, and now covered in trees. Near Cathiron, at grid reference SP 462785, a by-road crosses the old line of the canal close to a point where it diverges from the new; this bridge had recently been filled in, although the old canal beyond it, lined by willows, was still in water. It used to form a branch to Fennis Field lime works which the 1928 edition of *Bradshaw* notes as, even then, being 'not navigable'. On the other side of the filled-in bridge a reed bed indicated the line of the old canal, leading beneath a fine cast-iron roving bridge to give a glimpse of the new line beyond.

At grid reference SP 480774, another road crossed the old course, according to my 1967 edition Ordnance Survey map; but here, I found, it had recently been filled in. The old course here was part of a loop which almost, but not quite, rejoined the new line at Newbold on Avon. Newbold, with its old and new tunnels, and its canal pubs away from the canal, is the best known of locations where the new line and traces of the old may be seen. From the Braunston direction, the canal approaches on its original course, but then diverges to the north west into a cutting and tunnel. The old line contoured off to the south and the row of buildings which lines its route— now a gravel track—includes two pubs, the *Barley Mow* and the *Boat*. The latter is the farthest from the present canal, but close to the B4112 road, and to a casual passer-by offers no clue to the origin of its name. The old canal passed into a tunnel under the road and beneath the churchyard. A long dip across the lawn in front of the church marks its course. No trace is evident of the nearer portal, but the farther one remains complete in the corner of a meadow beside the church tower. Beyond it, a hedgerow, which now forms the boundary of the field, clearly follows the line of the towpath and must have been the canal's boundary hedge; a depression in the ground indicates the line of the canal.

This tunnel portal, mentioned in guides, is a popular canal sight, to judge from the succession of visitors, many dressed for boating, who appeared to inspect it while I was taking photographs. Some fifty yards or so of tunnel can be seen inside, complete; explorers on foot have to contend with its use as a shelter by cattle. Though why there was a tunnel here at all remains to me a mystery, for there is no great depth of earth above it.

The old tunnel, typical of its period, has no towing-path; the new tunnel, equally typical, has two. Since the canal otherwise has only one, each portal of the tunnel is in effect a turnover bridge with towpath carried

10/16 When the Oxford Canal improvements were completed in the 1830s, parts of the old line which served wharves were left in water as branches off the new direct line. This is Stretton Wharf, beside one of them, which is no longer navigable. (Grid reference SP 442810.)

up and over it. The consequence of the improvements generally was to leave the northern Oxford Canal as very much an 1830s canal, long flowing curves interspersed with straight sections, highish embankments, deepish cuttings, a pleasant iron aqueduct over a road near Rugby and this spacious tunnel which presented no hindrance to traffic. But every few miles the canal twists and turns along the contour for a few hundred yards where part of the old line was incorporated into the new.

The canal at Braunston is well known to canal enthusiasts, pleasure boaters and casual visitors. Yet the extent to which the layout of the canal is dependent on a history of improvements is not immediately obvious.

Here, in 1805, the wide Grand Junction Canal from London met the narrow Oxford Canal; narrow boats bound for Birmingham continued south and west along the Oxford Canal, following the contour for a few miles to Napton Junction where they diverged on to the Warwick & Napton Canal. The Grand Junction Canal joined the Oxford just north of the latter's Braunston Wharf, near the apex of a loop by which the Oxford Canal followed the contour. When improvements were made, a new route diverged from the old half a mile to the north. A gracefully laid-out triangular junction was made, leading boats from either direction on to the new line, which leads away into the distance, dead straight along a high embankment. Two cast-iron roving bridges linked by a brick arch on the central island of the triangle carry the towpath of the old line and a further turnover bridge a few yards down the new enabled horses to follow any possible permutation of route without obstruction. But this, for long the physical junction of routes, is called Braunston Turn: the junction between the canal companies remained at Braunston Wharf and half a mile of the Oxford Canal's old route remained in being as its Braunston branch, to give access to the Grand Junction and the wharf.

In 1929 the Grand Union Canal was formed, incorporating all the canals between London and Birmingham except the Oxford. However, as part of its scheme to widen the canals between Braunston and Birmingham, it was able to widen the Oxford Canal, where necessary, between Braunston and Napton. The principal work was re-

moval of the Oxford Canal's narrow stop lock, which was sited close to Braunston Wharf, just on the Birmingham side of the junction. Even now the canal here is less than full width, and the toll house still stands beside it. The origin of these features is not obvious, however, and to the observant boater a more immediately evident indication of moving from one canal to another, with neither curve nor physical change, is successive passage of bridge no. 91 (Oxford Canal) and bridge no. 1 (Grand Junction).

Prestolee and Ladyshore

The Manchester, Bolton & Bury Canal is generally little known to canal enthusiasts, for it is short, in an industrial area and, for the most part, was closed to navigation long before pleasure cruising became popular. So I am indebted to Bury Museum's *Local History Trail* booklet about the canal for pointing out how much of interest can still be seen.

The booklet is concerned with the Bury branch of the canal, most of which is still in water; it runs from Bury, Lancashire, in a south westerly direction for about four and a half miles to the top of Prestolee Locks (grid reference SD 752064). From this point, also called Nob End, the canal continued on the same level to the north west as the two-and-three-quarter-mile Bolton branch, while the main line dropped down the locks and headed for a junction with the Mersey & Irwell Navigation at Salford, eight miles away. Of these two lines, only short lengths in the vicinity of Prestolee, and another near Prestwich, remain in water.

The history of the canal was, briefly, this: it was opened in stages between 1796 and 1808, but its proprietors' early ambitions to link up with the Rochdale Canal (by extending the Bury branch) and/or the Leeds & Liverpool (by extending the Bolton branch) were thwarted and they became early supporters of railways. By 1837 they had converted their company into a canal and railway company, and had opened a railway between the Manchester and Bolton.

This followed a route distinct from that of the canal, which remained in use. By a series of amalgamations it became in due course the property of the London Midland & Scottish Railway and today it is one of BWB's remainder waterways, such of it as is left. For in 1936 there was a severe breach on the Bury branch near Prestolee, and subsequently the canal became by stages disused, unnavigable and abandoned.

Its hub was Prestolee and thither I went in March 1978. I approached the canal not through the village of Prestolee but from the north through Little Lever, bumping along an unsurfaced road lined with new houses, and then over a canal bridge to park at the water's edge on a wharf paved with cobblestones and still retaining its mooring rings.

From the bridge there was an extensive view of the scene, for the Bolton-Bury level of the canal is here carried high up near the rim of the Irwell Gorge. To the right the canal disappeared round a promontory in the direction of Bolton; in front traces of the old locks ran diagonally down the slope, at the foot of which a further section of canal, in water, crossed the Irwell by Prestolee Aqueduct; and to the left the canal was dry. But what at first appeared to be a small wood of willows was revealed by inspection as the site of an extensive basin, wide enough for boats entering it from the top lock to make a turn of almost 180 degrees back on themselves to pass beneath the bridge towards Bolton. Being laid out on a steep hillside, the canal had extensive brick and stone supporting walls on the downhill side, and it was only a few hundred yards towards Bury that the 1936 breach occurred.

The breach itself must have been spectacular, for the chasm that resulted could still be seen. It was never made good: instead, water was piped past the site. I could find no trace, however, of a boat which was left perched on the brink when the breach occurred. The site is, according to the Bury Museum booklet, unstable and to be visited, if at all, with the greatest caution.

The locks, long dry, led down the hillside in a long straight flight. There were two staircases, one after the other, of three locks each. Gates and paddle gear had gone, but most of the

10/17 (above) Remains of the upper of two staircases of three locks at Prestolee, Manchester, Bolton & Bury Canal, in 1978. Grid reference SD 752064.

10/18 (left) The same locks at the turn of the century, with a cotton-laden barge ascending.

10/19 (overleaf) Elegant cast-iron roving bridges dating from the early 1830s span the Oxford Canal at Braunston Turn. The original line is on the right; a cut-off, which avoided a long and winding section of contour canal, makes a triangular junction with it beneath the bridges.

chambers were still there, the canal bed a grassy slope between the walls.

They led down to another wide turning basin at the bottom, still in water (which is piped down from the upper section for industrial water supply). Here boats had to make another acute-angled turn to gain Prestolee Aqueduct. This was the most substantial surviving relic of the canal, a massive four-arched stone structure. A badly eroded path led down alongside the locks from wharf above to basin below. This no doubt was the route taken by travellers from the canal's passenger boats; before the railway was built, there was a thriving trade in passenger boats on this canal, but to save the time which would have been taken by working the boats through these locks, through passengers had to change boats, walking up or down them.

Before going to Prestolee I had visited Ladyshore, three quarters of a mile to the east. Here had been one of several collieries that were sunk close to the canal and provided it with much of its traffic. Ladyshore closed in 1951

and it and its surroundings have become a scene of real industrial dereliction, bleaker than ever on a March day when a strong cold wind brought showers of chilling rain. It was difficult to appreciate that there had once been a thriving enterprise here which built and repaired its own boats at the canal side.

Fortunately, from the point of view of the canal archaeologist, the influence of the redeveloper has not yet been felt at Ladyshore, nor at Prestolee, and at both places many traces of the past remain to be seen. At Ladyshore the canal was in water and there was a nice milestone, real stone, beside it. But the main interest of the place lay in the sunken remains of several boats. These were of unusual pattern, narrow boats slightly shorter than usual (the locks were built for boats sixty eight feet long) which carried coal in small wooden containers. These were lifted in and out by crane. By walking along the towpath, across the dam which closes off the drained section of the canal past the breach, and back on the offside, it was

possible to examine these at grid reference SD 759066. Enough of one of them remained above the surface to identify it, while beneath the ripples the row of containers could still be discerned in its hold.

Bingley

The chambers of Bingley Five-Rise Locks on the Leeds & Liverpool Canal are wide but short, after the manner of northern canals—they take boats no more than sixty two feet long. So this staircase of five locks rises steeply and impressively. Its vertical ascent is sixty feet.

The staircase is located at grid reference SE 107399. The easiest road approach is to the top, from which there is a good panorama of the mill chimneys and moors of West Yorkshire; but the locks are seen at their most impressive from below. Deep sombre black gates alternate up the hill with spindly white wooden bridges across each chamber, and stone retaining walls rise beside the flight like ramparts.

The lock chambers themselves are built from large blocks of stone and alongside at each level is ground paddle gear of a type unfamiliar to those who know only southern canals. It is operated by winding a large handle horizontally. There are gate paddles too and at least one pair of gates has paddles which are pivoted to slide through an arc sideways to open their sluices. Plaques at the top of the flight record its construction in 1774 by four local masons to the design of John Longbotham of Halifax.

There is a modern swing bridge above the top lock, and a much older one above the top lock of Bingley Three-Rise Locks. This staircase is a few hundred yards in the direction of Leeds and, massive and impressive, would be famous in its own right if it were not overlooked by the Five-Rise staircase. Visiting the area in March 1978 I found it well worthwhile to walk down to them and then on for a long mile to Dowley Gap Aqueduct. Beside Bingley Three-Rise Locks stood an extensive stone-built mill which produces warm underwear of a brand sometimes advertised in the boating press, although no mooring was provided for passing boaters to call. By

10/20 Bingley Five-Rise Staircase, Leeds & Liverpool Canal. With lock chambers only a little over sixty feet long, the staircase rises with dramatic steepness.

10/21 Grand Canal, 11th Lock. Work on this section of the canal commenced under Thomas Omer as early as 1756. The wide part of the lock chamber is Omer's work; narrows at the gates were added later to conform with other locks of smaller dimensions recommended by Smeaton. Remains of Omer's lock cottage are in the background; the straightness of this canal is in marked contrast with winding contour canals built later in England.

contrast a timber merchant's establishment a little farther on had adapted its wharf for pleasure craft and invited visitors to inspect its do-it-yourself section. Adjacent was another mill, or factory, its boiler room facing the canal with open furnaces glowing dimly red as the fires were settled down for the night, and large piles of fuel for them were spread out along what was once a wharf. No doubt this mill was laid out to make use of canal transport, as certainly was a coal yard nearby, this time on the towpath side of the canal, which still retained an electrically operated grabbing crane to unload boats.

Town and country intermingle in these parts, and by Dowley Gap two-lock staircase the canal was rural again. Immediately west of the locks was bridge no. 205, one of those bridges of which the stonework suffered so much erosion by tow-ropes that a wooden roller was fitted to keep

ropes off the edge of its arch. To judge from the absence of scoring on the roller, it must have been installed fairly recently, by canal standards.

Close beyond is Dowley Gap Aqueduct, stone built, wide, low and long (but not squat), carrying the canal across the River Aire in a tree-shaded glen. Between locks and aqueduct is Dowley Gap Changeline Bridge, a turnover bridge by which the towpath changes sides, with the usual steeply curving pathway on one side for horses, having crossed over the canal, to pass beneath the arch without detaching tow-rope. Where a canal runs along the side of a valley, the towpath was generally built along the downhill side which had to be embanked anyway: at Dowley Gap the canal crosses from one side of the Aire valley to the other, hence the turnover bridge. In this instance the builders economised by utilising a bridge which also carries a lane.

The Grand Canal at Dublin

Near Clondalkin the Grand Canal, which has crossed Ireland from the Shannon, approaches Dublin; there are still fields nearby but housing estates appear in the distance. The canal is popular with their young inhabitants for fishing and also, since the water is exceptionally clear, bathing. Below 11th Lock swimmers are particularly fortunate, for at the tail of the lock, for twenty yards or so, the edges of the canal are built up with masonry and these walls, wider apart than the lock entrance but narrower than the usual width of the canal, turn the tail of the lock into a convenient swimming pool.

To the canal archaeologist, however, these and other features of 11th Lock are of much greater interest, for this was the first lock to be built on the Grand Canal, and work started as early as 1756 or thereabouts. The lock is not in its original condition but much original work remains.

The engineer in charge, Thomas Omer, started work near Clondalkin to avoid the high cost of land nearer

Dublin. Between 1756 and 1763 he constructed several miles of canal and three locks, which eventually became the 11th, 12th and 13th Locks (numbered from Dublin) when the canal was extended. He built, however, for a truly grand canal: dimensions of lock chambers were no less than 137 feet by 20 feet. Later engineers were more modest: Trail, engineer to the Grand Canal company, reduced the lock dimensions to 80 feet by 16 feet when he recommenced work on the locks in March 1773 with 1st Lock, in Dublin. But John Smeaton, called in as a consultant later the same year, recommended that even this was too large. Taking into account the likely extent of trade, he recommended locks only 60 feet long by 14 feet wide: boats able to pass through these would carry about 40 tons and would be little delayed waiting for full loads. These dimensions were adopted as standard.

Omer's original locks were altered to conform. At 11th Lock this meant that new gates, of standard width, were installed, the standard length apart. For these, the lock chamber was narrowed, but only near the gates: most of the existing lock chamber is wide, and the masonry of the wide part is Omer's work. This appears to be the oldest masonry still in use in a canal lock in the British Isles.

The original chamber walls still extend for a short distance above the lock, and a much longer distance below it. There, however, they have been lowered so that they rise only a foot or two above the water's surface. They provide swimmers, unaware of their historic significance, with a convenient way to enter the water—the edges of the canal elsewhere are covered with reeds and bushes. In these walls may be clearly discerned the recesses for the lower gates of Omer's lock. The recesses for the top gates appear to have been obliterated in rebuilding for narrow gates.

The 12th and 13th Locks also have similarly wide chambers with narrow gates. At 12th Lock the remains of the wide lock extend below the existing lock, but 13th Lock was later converted into a double lock or staircase— the upper of its two chambers is wide and the rest of Omer's lock is replaced by the lower chamber. Also to be seen at 11th and 12th Locks are handsome little lock houses constructed by Omer. The one at 11th Lock is ruined but still

substantial, while the one at 12th Lock is inhabited. For some reason no longer clear, these houses were built not beside the locks but on the canal bank a few hundred yards above them.

In Dublin, below 1st Lock (which is a double lock and retains Trail's larger-than-standard chambers) the canal, which has run straight for miles, turns sharply to the right. This was, until recently, a junction, with the main line running straight ahead for a mile to James's Street Harbour, the canal's original Dublin terminus. Now the section to the harbour is filled in and landscaped: grass and trees have been planted and footpaths laid out. Its sole overbridge now spans a footpath.

When I visited James's Street Harbour in 1969 I was able to arrive by boat. On my return, nine years later, inevitably by car, I drove up to find the surroundings little changed: the same cobbled quay, the same row of canal houses with the canal manager's house, more ornate than the others, prominent mid-way along it, the same dilapidated sheds over the dry docks (one of them had, by my return visit, actually collapsed). But of water there was no trace: beyond the edge of the quay with its mooring rings, all was filled in, waste ground covered with weeds, separated from the quay by a wire fence through which enterprising spirits had forced their own gaps. It was all most confusing, almost eerie.

The sign above the road entrance which used to announce *CIE Canal Section* had also gone, but many canal warehouses still stood—adapted to other uses—and the Harbour Lights Bar was still doing business in Grand Canal Place, which is now over a mile of city from the nearest part of the canal in water.

So below 1st Lock the continuation of the canal through Dublin is not its Main Line, but its Circular Line, built between 1790 and 1796 over a roughly semi-circular course to link the canal with the River Liffey and the sea at Ringsend, about four miles long with seven locks. Counting from Ringsend up the canal, therefore, the lock known as 1st Lock is in fact the eighth.

The Circular Line is level for a couple of miles to Portobello, where La Touche Bridge carries Richmond Street over the canal. In 1805 the board of the canal company decided to make the terminus of the passage boats at

Portobello and to build a hotel for passengers. The building, which is still prominent beside the canal, is a handsome one: it was completed and opened as a hotel in 1807 and continued in use as such under various tenants until about 1855. It must have been an attractive place to stay: the peace and quiet of the tree-lined waterway enlivened from time to time by the bustle of arriving and departing passage boats.

Since closure as a hotel the building had a succession of uses. It is now used as offices and happily the recent change has resulted in sympathetic restoration externally as has been described in chapter nine.

Canal basins which once opened off the canal at Portobello have long since been filled in; the car park beside the hotel building marks the site of one of them, and an adjacent factory another.

From Portobello almost to Ringsend the Circular Line falls by a series of short straight pounds interspersed by locks: at each lock the route bends slightly to the left, and at its tail one of the city streets is carried over the canal. This section is most attractive: unlike the hidden canals of Birmingham, Dublin's canal was laid out to adorn the Georgian city: it is lined by lawns and trees, and eighteenth-century terraces. Huband Bridge (Upper Mount Street) is more ornate than most—it was ornamented and balustraded at his own expense by Joseph Huband, a director of the company.

The docks in which the canal terminates at Ringsend seem enormous, but have never been a commercial success and are now little used. Three parallel entrance locks from the Liffey, all dating from the 1790s, indicate the anticipated scale of traffic; the smallest (barge size) and the largest (for ships) remain usable.

The continued existence of the Circular Line has been seriously threatened on at least two occasions. In the 1860s the canal company agreed to sell it for conversion to a railway— fortunately the sale was never completed. A century later, in the 1960s, proposals were made to use its course for a sewer and a motorway. Opposition to the latter proposals, led by the Inland Waterways Association of Ireland, was successful and happily the Circular Line is still extant for the enjoyment of both Dubliners and canal enthusiasts.

The Union Canal and its aqueducts

If Brindley's early canals are likened, in their stage of development, to a veteran car of the most primitive sort, then the Edinburgh & Glasgow Union Canal is a veritable vintage Bentley. It demonstrates all the confidence, the panache, the ability to combine elegance with utility, that canal engineers had gained with sixty years of experience.

It was laid out from Edinburgh to Falkirk, for thirty one miles as a single long pound, followed by a single flight of eleven locks down to the Forth & Clyde; and it was opened in 1822. It could not, however, achieve that long pound simply by winding along the contour. Certainly, there are sections where it does run along hillsides—and some substantial embanking on the downhill side of the canal is needed along some of them. But elsewhere embankments lead the canal across valleys by sweeping curves, and rock cuttings are carved through higher ground. There is at Falkirk a 696-yard tunnel, with towpath, through rock—the only canal tunnel in Scotland of any great length—and across valleys too deep for embankments there are three magnificent aqueducts, second only to Pontcysyllte for grandeur.

Elsewhere, although humped canal bridges of stone come frequently, some of them with balustrades of iron, small aqueducts over roads and lanes are as frequent, as common as underbridges on a railway. Even the humped bridges are unusual in their context, for alone among surviving Scottish canals the Union Canal is of purely inland significance. The others are, or were, sea-to-sea links, and so have, or had, opening bridges for ships with masts.

The Union Canal saw only a short period of full commercial use, for the

10/22 (above) The arches of Avon Aqueduct, Union Canal, silhouetted against the valley floor. Grid reference NS 967758.

10/23 (below left) The majestic proportions of Avon Aqueduct are best appreciated from below.

10/24 Causewayend Basin (grid reference NS 962762) was built for exchange of passengers, as well as freight, between trains of the Slamannan Railway and boats on the Union Canal, which passes the entrance in the background.

10/25 (right) Culverted road crossings mark the position of former bridges on many closed canals. This was Preston Road Bridge, Linlithgow, on the Union Canal. Participants in the 1978 Glasgow-to-Edinburgh inflatable boat marathon are lifting their boats from the water, to re-launch beyond the obstruction.

Edinburgh & Glasgow Railway was opened twenty years after the canal. The latter passed eventually into railway ownership; commercial traffic died out in the early 1930s and the locks at Falkirk were filled in. So was Port Hopetoun, the main basin in Edinburgh, but between these points the canal remained intact for another thirty years and in 1959 British Waterways were enterprising enough to put two small hire cruisers on it. These, ahead of their time, were unsuccessful, and in 1965 the canal was closed to navigation.

Since then the waterway has been split up into ten separate lengths by obstructions such as culverted bridges, a motorway crossing and a half-mile length filled-in and piped in connection with a new housing estate.

It was little less than tragic that, after surviving almost intact for so long, the canal was closed and split up at that date: for not only did it have to continue to exist as a watercourse to supply industries with water in a district where water is often in short supply, but within a few years there was an outburst of interest in its value as an amenity. When I first came across the Union Canal, and marvelled at it, in 1970, it was little known; since that date the Scottish Inland Waterways Association has been formed, largely as a result of concern over the canal; numerous local canal societies along its route have followed; boating has been reintroduced (including an excellent restaurant boat, *Pride of the Union*, based at Ratho); and the canal and its future have been the subjects of investigations and reports, official and unofficial, almost ad nauseam. But the reports have generally been favourable, and the outlook, despite proposals for another low-level road crossing, is brighter than it was.

Many of the canal's bridges and other structures have been protected by scheduling as ancient monuments or listing as buildings of historic interest—presumably at the behest of the same local authorities, or their successors, which were earlier all for demolishing bridges and replacing them by culverts.

The outstanding feature of the canal is its three great aqueducts. The greatest of these is Avon Aqueduct (grid reference NS 967758), which is 810 feet long, 86 feet high at the most, and has twelve arches: figures exceeded only by Pontcysyllte Aqueduct. Almond Aqueduct (grid reference NT 105706) is 420 feet long and 76 feet high with five arches, and Slateford Aqueduct (grid reference NT 220707), over the Water of Leith in the south western suburbs of Edinburgh, is 500 feet long, 75 feet high and has eight arches. These aqueducts were not built for narrow boats but for barges 12 feet 6 inches wide. All three aqueducts are of similar design, by canal engineer Hugh Baird; they resemble not Pontcysyllte but Chirk, as rebuilt—that is to say, a masonry structure with an iron trough channel. The piers are farther apart than those of Chirk, and the arches of greater radius. This radius is also deep in relation to the overall height of the aqueducts: these features combined make it difficult to give a clear impression of the scale of these structures in photographs, and personal visits were, for me, a revelation.

In the vicinity of Avon Aqueduct are several other features worth visiting. A few hundred yards to the west is Causewayend Basin (grid reference NS 962762). This rectangular inlet leading off the canal is one of the few places where a railway basin was used for interchange not only of freight, but also of passengers. Opening of the Slamannan Railway in 1840 completed a rail link from Glasgow to Causewayend, and it was the fond hope of the Union Canal company that the railway-canal route would become the principal passenger route between Glasgow and Edinburgh. It was a hope in vain, for not only was a coach service put on in competition with the canal, but two years later the Edinburgh & Glasgow Railway was opened.

The Union Canal is particularly rich in relics of passenger-carrying; I have already mentioned in chapter nine its westernmost extension to Port Maxwell for passenger traffic, and the range of stables built for passenger-boat horses at Woodcockdale. The latter place is about half-a-mile east of Avon

Aqueduct. A couple of miles further on is Linlithgow. Here the Linlithgow Union Canal Society has established a small but worthwhile canal museum in another range of old canal stables. It also runs passenger trips in its boat *Victoria*, although this cannot go to the aqueduct for the way is blocked by the culverted Preston Road Bridge (grid reference NS 997765).

In the opposite direction, on the outskirts of Linlithgow, the canal crosses over the A9 main road by a good example of an underbridge or aqueduct—call it what you will. The roadway beneath it is, by modern standards, narrow, and the obstruction that the canal at this point causes to road traffic was one of the arguments put forward in favour of the 1965 closure. But through traffic now bypasses Linlithgow by motorway, the aqueduct is still there, and the canal.

The continued existence against all obstacles of the Union Canal, even as the Union in fragments, seems an appropriate note on which to end. For old canals and their artefacts are persistent things: they were built to last

and they do. They serve as a reminder that the present day, for all its technological achievements, has no monopoly of skills—and furthermore that things which today seem most irreplaceable (airports, perhaps, and motorways) may prove to have as transient a useful life. They in turn will leave traces of their own past for future generations to marvel over. So far as the canals of the canal era are concerned, there is no doubt that much will remain for a very long time to come to fascinate the canal archaeologist. What is important, however, is that as many canals as possible should remain navigable, with their historic features carefully maintained. Margaret Drabble, experiencing a canal holiday recently for the benefit of readers of the *Telegraph Sunday Magazine*, headed her article *A Boat Ride back in Time*. And that is the point: it is possible, of course, and has a certain fascination, to interpret the past from the dead remains of closed canals—but how very much better to be able to experience it oneself on a canal which remains in use and alive.

ACKNOWLEDGEMENTS

A book of this nature can be written only with the help
of many people, and to all of them I am most grateful.
The following have been particularly helpful in
searching out answers to my sometimes abstruse
questions:
J.K. Allan (Falkirk Museums): N.S.G. Bostock;
J.C. Brown, Philip Daniell, Sheila Doeg (British
Waterways Board); Tony Conder (Waterways
Museum, Stoke Bruerne); Mrs Ruth Heard; Tony
Hirst (Boat Museum); A.D. Hodge (Manchester Ship
Canal Co.); K. Howarth; T. Insull (Cadbury
Schweppes Ltd); Hugh Malet; D.L. McDougall
(Black Country Museum); Kevin Raynor (Arthur
Guinness Son & Co. Ltd); Michael Streat; Mrs Helen
Theakston;
and the staffs of: Lincolnshire Central Library;
Lincolnshire County Archives; National Library of
Ireland; National Maritime Museum; North
Yorkshire County Archives.

Canalside Museums

Waterways Museum
(Stoke Bruerne, Northants)
Definitive collection of canal artefacts.

Boat Museum
(Ellesmere Port, Cheshire)
Unique and excellent collection of canal boats and barges.

Black Country Museum
(Dudley, W.Midlands)
Incorporates branch off Dudley Canal with boat dock, limekilns etc.

Blists Hill Open Air Museum
(Ironbridge, Shropshire)
Incorporates tub-boat canal with inclined plane.

Cheddleton Flint Mill
(Cheddleton, Staffordshire)
Incorporates canal wharf and preserved narrow boat.

Exeter Maritime Museum
Mainly sailing vessels, but housed at Exeter canal basin.

Linlithgow Union Canal Society Museum
Small but worthwhile museum relating to Union and Forth & Clyde Canals.

Llangollen Canal Exhibition
Story of canals vividly told by audio-visual means in old canal warehouse.

Robertstown
Museum of canal artefact and other exhibits of local interest, housed in former hotel for canal passengers.

SELECT BIBLIOGRAPHY

History

The Canals of the British Isles series (Newton Abbot: David & Charles)

Boyes, J. & Russell, R. *The Canals of Eastern England* 1977

Delany, V.T.H. & D.R. *The Canals of the South of Ireland* 1966

Hadfield, C. *The Canals of the East Midlands* 1966

Hadfield, C. *The Canals of the West Midlands* 1966

Hadfield, C. *Thé Canals of South and South East England* 1969

Hadfield, C. *The Canals of South Wales and the Border* 1960

Hadfield, C. *The Canals of Yorkshire and North East England* 1972

Hadfield, C. *The Canals of South West England* 1967

Hadfield, C. & Biddle, G. *The Canals of North West England* 1970

Hadfield, C. & Norris, J. *Waterways to Stratford* 1962

Lindsay, J. *The Canals of Scotland* 1968

McCutcheon, W.A. *The Canals of the North of Ireland* 1965

Rolt, L.T.C. *From Sea to Sea - The Canal du Midi* (London: Allen Lane 1973)

Rolt, L.T.C. *Thomas Telford* (London: Longmans Green & Co. 1958)

Malet, H. *Bridgewater The Canal Duke, 1736-1803* (Manchester: Manchester University Press, 1977)

Delany, R. *The Grand Canal of Ireland* (Newton Abbot: David & Charles 1973)

Ellis, H. *British Railway History* (London: George Allen & Unwin Ltd, Vol. I 1954, Vol II 1959)

Spratt, H.P. *Science Museum Handbook of the Collections Illustrating Marine Engineering* (London: H.M.S.O. 1953)

Whitwell, J.B. *Roman Lincolnshire* (Lincoln: Lincs Local History Society, 1970)

Gore, D., Greenhalgh, J. & Smith, C. *The Lincolnshire Car Dyke, A Field Survey* (Nottingham: University of Nottingham, 1970)

Guides

Langford, J.I. *A Towpath Guide to the Staffordshire & Worcestershire Canal* (Cambridge:Goose & Son Publishers Ltd, 1974)

Stevens, R.A. *A Towpath Guide to the Brecknock & Abergavenny and Monmouthshire Canals* (Cambridge: Goose & Son Publishers Ltd, 1974)

Edwards, L.A. *Inland Waterways of Great Britain* (St Ives, Huntingdon: Imray, Laurie, Norie and Wilson 1972)

Sullivan, Dick *Old Ships, Boats and Maritime Museums* (London: Coracle Books, 1978)

Atterbury, P. and Darwin A. (editors) *Nicholson's Guides to the Waterways* (vols 1 to 4) (London: British Waterways Board, undated)

General

Lewery, A.J. *Narrow Boat Painting* (Newton Abbot: David & Charles 1974)

Chaplin, T. *A Short History of the Narrow Boat* (Shepperton: Hugh McKnight Publications, 1974)

Booklets in a series 'covering all aspects of boating on inland waterways' (Kettering: Robert Wilson):

Wilson, R.J. *The Number Ones* 1972

Faulkner, A.H. *The George and The Mary* 1973

Paget-Tomlinson, E. *Mersey and Weaver Flats* 1973

Wilson, R.J. *Knobsticks* 1974

Wilson, R.J. *Boatyards and Boatbuilding* 1974

Faulkner, A.H. *FMC* 1975

Fraenkel, P. and Partners *The Waterways of the British Waterways Board A Study of Operating and Maintenance Costs* (London: Department of the Environment 1977)

Fourth Report from the Select Committee on Nationalised Industries, Session 1977-78, British Waterways Board (London: H.M.S.O. 1978)

Periodicals

Waterways World (Burton-on-Trent: Waterway Productions Ltd)

Industrial Past (Skipton: John Keavey)